Mahamud Subir

Colloidal Interfaces

Also of Interest

Polymers, Colloids, and Surface Chemistry.
Polymerization, Suspensions, Emulsions, and Adsorption Fundamentals
Edited by: Shramila Yadav and Meenakshi Gupta, 2026
ISBN 978-3-11-914664-7, e-ISBN (PDF) 978-3-11-220823-6

Quantum Mechanics
Guo-Ping Zhang, Mingsu Si and Thomas F. George, 2024
ISBN: 978-3-11-067212-1, e-ISBN (PDF) 978-3-11-067215-2

Carbon for Micro and Nano Devices
Swati Sharma, 2024
ISBN: 978-3-11-062062-7, e-ISBN (PDF) 978-3-11-062063-4

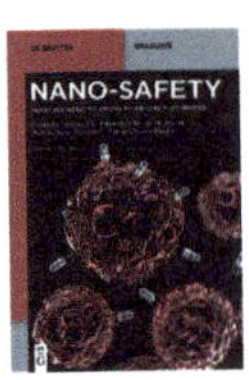

Nano-safety.
What We Need to Know to Protect Workers
Edited by: Dominick E. Fazarro, Christie M. Sayes, Walt Trybula, Jitendra S. Tate and Craig Hanks, 2023
ISBN 978-3-11-078182-3, e-ISBN (PDF) 978-3-11-078183-0

Non-equilibrium Thermodynamics.
Entropy, Heat and Mass Transfer
Michael Bestehorn, 2024
ISBN 978-3-11-079404-5, e-ISBN (PDF) 978-3-11-079413-7

Mahamud Subir

Colloidal Interfaces

Laser Characterization

DE GRUYTER

Author
Prof. Mahamud Subir
Department of Chemistry
Ball State University
Muncie
47306 Indiana
USA
msubir@bsu.edu

ISBN 978-3-11-101819-5
ISBN 978-3-11-101838-6 (PDF)
ISBN 978-3-11-101893-5 (EPUB)
DOI https://doi.org/10.1515/9783111018386

Library of Congress Control Number: 2026932519

Bibliographic information published by the Deutsche Nationalbibliothek
The Deutsche Nationalbibliothek lists this publication in the Deutsche Nationalbibliografie; detailed bibliographic data are available on the internet at http://dnb.dnb.de.

De Gruyter and Walter de Gruyter GmbH are part of De Gruyter Brill.
www.degruyterbrill.com

Questions about General Product Safety Regulation:
productsafety@degruyterbrill.com

Cover illustration: Moussa81/iStock/Getty Images Plus
Printing: CPI books GmbH, Leck

To my parents, Nazma and Arun, who made all of this possible.

Acknowledgments

Special thanks to Prof. Yi Rao for his critical review of the manuscript.

 | https://doi.org/10.1515/9783111018386-202

Preface

Chemistry topics involving gaseous and solution phases are commonplace in most undergraduate chemistry curricula. The foundational – general, organic, inorganic, analytical, physical, and biochemistry – courses predominantly focus on chemical processes that take place in the bulk, homogeneous medium. However, the natural world and chemical advances that move society forward include complex, heterogeneous, multi-phase systems. One such example is the colloidal state of matter. Colloids are chemical and physical systems comprised of small dimensions and large surfaces. In other words, colloids are particulate matter, one bulk medium (solid, liquid, or gas) dispersed in another. Naturally, then, colloids involve a molecular boundary between two distinct phases, which we call an interface. Thus, the study of colloids is inexorably linked to the investigation of surface or interfacial chemistry. Understanding fundamental aspects of colloids and their interfaces, and application of lasers to study them in situ is the focus of this book.

Colloids are tiny particles, in the range of nano- and micro-meter size. The interface, which is buried between two phases, is in molecular dimension. Experimentally, studying colloids and colloidal interfaces is therefore not trivial. One way to probe colloids and their interfacial properties is based on light. Photons bouncing off particles can contain information about their physical and chemical attributes. Analyzing these detected lights – their intensity, scattering direction, polarization, etc. – can reveal many physicochemical properties of dispersed particles and molecular behavior at colloidal interfaces. With the advent of laser, investigating physical characteristics of colloids, such as their size distribution and charge properties, has become routine, yet not without challenges. In recent decades, the application of high-intensity laser has made it possible to extract molecular-level insights into interfacial chemistry of various types of colloids. Namely, the field of interfacial selective nonlinear optics (NLO) has paved the way to explore colloidal interfaces in a way that has not been done before. New pictures are still emerging. The basics of these laser light scattering techniques are integral topics of this STEM book.

There are many excellent books on colloidal and interfacial science. So, how is this book unique? There is a confluence of reasons for writing this book. The five-volume series on *Fundamentals of Interface and Colloid Science* by J. Lyklema and *Principles of Colloids and Surface Chemistry* by P. Hiemenz and R. Rajagopalan are a must-read for those fully immersed in colloidal and interfacial research. These books explore a broader scope of colloidal chemistry (synthesis, design, stability, thermodynamics, application, etc.) and are mathematically detailed on various aspects of colloids and interfacial processes. They can be overwhelming for the uninitiated. Recent advances on colloidal interfacial spectroscopy are absent from these books. In contrast, this condensed STEM book, with its narrow focus on interfacial chemistry and application of lasers in colloids and colloidal interfaces, is geared toward nonspecialists as well as seasoned scientists with expertise in other areas of chemistry. A direct

 | https://doi.org/10.1515/9783111018386-203

connection between the applications of lasers and investigation of colloids is made in this book. New developments in laser spectroscopy for the investigation of colloidal interfaces is an integral component.

Employing pedagogical approaches (illustrations, simulations, examples, and chapter-end problems), this book offers a review on the major theories pertinent to colloids and their surface properties for students and career starters. As alluded at the onset of this preface, most undergraduate chemistry curricula do not explore colloid and interfacial science in detail. Thus, this book can serve as a steppingstone for curious minds interested in a deep dive into the world of surface and colloidal research. Throughout this book, we shall see that colloidal chemistry is truly interdisciplinary. This book takes the well-established basic principles of colloidal science and connects them to the up-and-coming advanced laser-based research on colloidal interface. Contemporary research involving colloidal interfacial chemistry is showcased. Accordingly, it is the author's intention to inform and inspire researchers in different scientific areas and to provide the necessary toolsets to implement the fundamentals of colloids and interfaces and broaden the horizon of this interdisciplinary research.

We shall begin (Chapter 1) with a survey of colloids in nature. Their application in our society will highlight the importance of studying colloids and colloidal interfaces. Colloids are small, but their impact is colossal. The discussion will include a brief historical overview, capturing the evolution of colloidal science in time. In this journey, the readers will get a glimpse of how extensive colloidal science is. The intermingling of theory and experiments that shaped the field of colloidal science will become evident. Chapters 2 and 3 are dedicated to reviewing the rudimentary physical, chemical, and electrical properties of colloids. Classifications, definitions, and terminologies associated with colloids and interfacial science are presented in these chapters. Adsorption is a key aspect of interfacial chemistry and is discussed in Chapter 3. Chapter 4 serves as a bridge, as it provides an overview of major experimental techniques relevant to studying colloids and interfacial chemistry. Properties of light and, in particular, lasers are discussed in the context of measuring inherent properties of colloidal systems. It will be apparent that light scattering techniques, the focus of this book, are essential in studying colloids.

Chapter 5 presents classic examples of static and dynamic light scattering methods to study colloidal properties such as size, zeta potential, and molecular weight. A powerful tool to investigate colloidal interfaces in situ is NLO laser spectroscopy. Chapter 6 delves into the basic tenets of these interfacial selective techniques with emphasis on second harmonic and sum frequency generations. Research findings and future direction involving colloidal interfaces, as probed using laser-based nonlinear spectroscopy, are reviewed in Chapter 7. An outlook on the future of colloidal interfacial science is provided in the Epilogue.

Contents

1 Colloidal and Interfacial Science: A Journey Through Time

1.1 Colloids in Nature and Society

What are colloids and why are they important? The importance of colloids is easy to see, as they are everywhere. It is more than likely that all of us have seen, touched, tasted, and even breathed in colloids. If you ever strolled amidst a foggy morning and felt the mist on your face, you have seen and felt colloids. Fog and mist, differing only by their density and location, are colloids. They are minute droplets of liquid water dispersed in the gaseous phase such as air. Light scatters off these suspended particles, allowing us to see them – a phenomenon known as the Tyndall effect. Tasting colloids is perhaps more conspicuous. Albeit different types, milk, butter, mayonnaise, and Jell-O are all examples of colloids. Butter, for instance, is liquid water dispersed in solid fat – a type of gel or solid emulsion. As for olfactory sensation, pump a bottle of perfume and take a whiff of the spray, there you have it! The aerosols generated are colloidal liquids suspended in air.

Given their ubiquity, understanding the physicochemical properties of colloids and chemistry at their interfaces (the spatial boundary between the dispersed and the continuous phases) is of paramount importance. Environmental processes and technological advances depend on colloidal chemistry. Aerosols in the atmosphere are minute airborne particles of solids and liquids. They originate from both anthropogenic and natural sources. Dust, sea salt spray, smoke from fires, fog, and mist are some examples of atmospheric aerosols. These fine particles, not seen to the naked eye, play an important role in radiative processes and impact earth's climate. Depending on their size and composition, they can scatter or absorb sunlight and heat. They can have cooling or heating effects. Atmospheric organic and bioaerosols, which are aerosols of biological origin, have been shown to serve as cloud condensation nuclei, which lead to the formation of cloud droplets and ice crystals. Aerosols containing toxic compounds are also of health concern.

Figure 1.1 displays examples of colloids found in the environment and synthesized colloidal materials important in technology. Colloids are abundant in the aquatic environment as well. Natural waters contain an assortment of dissolved macromolecules and suspended fine particles. They are typically classified as particulate organic matter and dissolved organic matter. These terminologies are operationally defined based on the particles' ability to pass through a 450 μm mesh filter. The general depiction of aquatic natural organic matters (NOMs) is that they are gel-like polymers exhibiting colloidal behavior. NOMs have a wide variety of compositions and sizes [1–6]. Humic acid and fulvic acid, two major macromolecular components of NOM, can range from 1 to 100 nm, whereas their aggregates can be in the micron range. These particles, along with suspension of clays, metal oxides, and soil aggregates can have sufficient

 | https://doi.org/10.1515/9783111018386-001

sorption capacities and thereby alter the fate and transport properties of aquatic pollutants. Man-made colloids are also detected in the marine environment and in freshwater. In recent years, concerns have been raised about plastic contaminants. They enter the environment and break down via physical and chemical processes. Nano- to micron-sized plastics have deleterious effect on aquatic organisms, and they also serve as vectors for transporting organic pollutants [7–9]. Their physicochemical behavior can be understood based on colloidal and interfacial science [10].

One thing apparent from these examples is that colloids are small, in the nano- to micrometer range. Their dimension is between that of molecules and macroscopic objects. Figure 1.2 puts the colloidal dimension in perspective. It is difficult to draw a distinct line between colloids, molecules, and suspensions. Suspensions are large solid particles dispersed in a liquid and can be seen with the naked eye, whereas molecules are not visible. Depending on particle density, a colloidal dispersion may appear translucent or opaque, i.e., they scatter light. At this length scale, surface chemistry also becomes important. Because these particles are small, they have a great deal of interfacial regions per unit mass. As an example, let us consider a Rubik's cube – a chunk of solid plastic that is cubic in shape. With a length of 5.75 cm, it occupies a volume of ~190 cm^3. The surface area of this piece of plastic is ~200 cm^2. Imagine that we break it evenly into quadrillion (10^{15}) pieces. Each particle now has a length of roughly 0.6 μm and a volume in the order of 10^{-19} m^3. This size is within the colloidal range and the total volume, whether it is broken down or intact, remains the same. The total surface area of all these minute particles, however, is ~2,000 m^2. This is bigger than the area of an Olympic-size swimming pool! Essentially, the smaller the matter, the more of its surface is exposed to the surrounding, and is available to interact and react with surrounding chemical species. Many applications, e.g., catalysis and nanotechnology, take advantage of this inherent interfacial property of colloids. In the grand scale of things (Figure 1.2), colloids are insignificant in size and occupy a narrow window in the spatial domain, but their impact on our society is gigantic.

Humans produce a variety of colloids, and they are used in our daily lives. Paints and inks are prime examples of commercial colloids. Paint is a colloidal dispersion of dried pigment nanoparticles, which can be of different inorganic or organic materials, such as TiO_2, chalk, gypsum, silicates, and latex beads [11]. They can come from metals, plants, and even insects. They are dispersed in a medium that is either aqueous or nonaqueous solvents or diluents (thinners). Another component of paint is the binder, which forms a continuous film on the surface of the substrate. As the name suggests, it holds or binds the pigment particles together, forming a coating on the substrate. Polymeric resins are common types of binders, which are dispersed in the solvent either as colloids or a true solution. Similarly, ink is a colloidal mixture of wax, resin, dryer, reducer, and organic pigment that gives the color. In olden days, ink makers used soot as a pigment for black ink. Pigments are dispersible particles that impart color. The pigmented particles used in ink and paint shed light on the adsorptive properties of various colloids. As such, colloidal systems are often applied in adsorp-

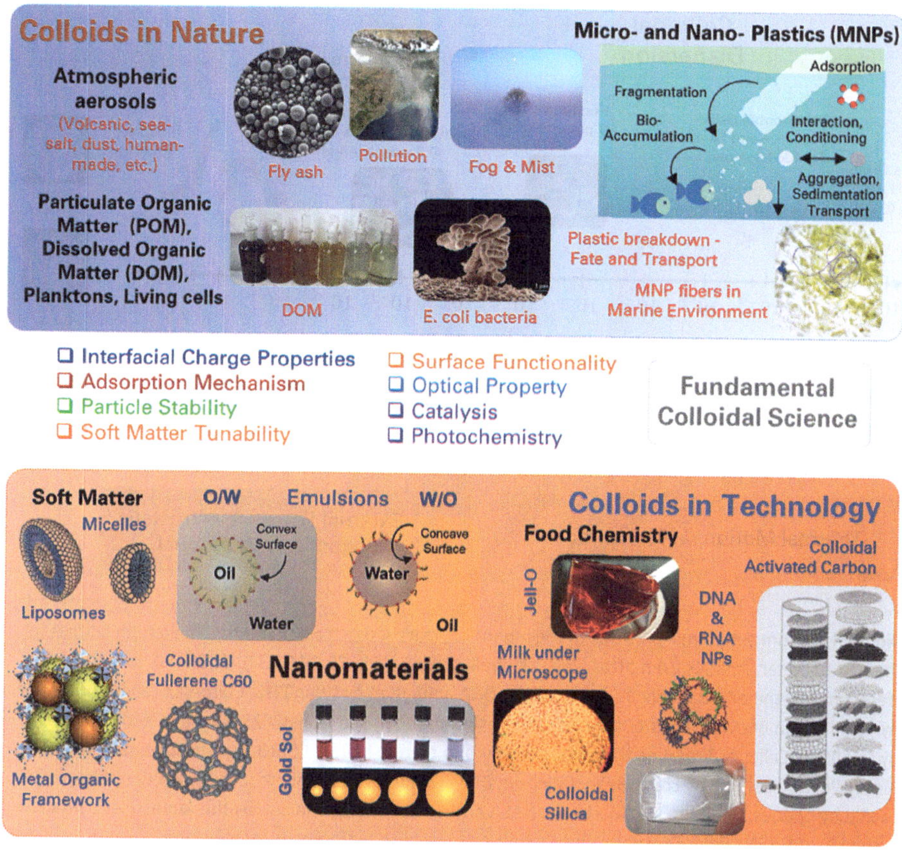

Figure 1.1: Colloids in nature and applied science. A pictorial representation of different types of colloids found in the environment and used in technology. Aspects of fundamental colloidal science are listed. Images from https://en.wikipedia.org/wiki/ via CC BY-SA 3.0, CC BY 4.0. No changes were made to the original images. Attributions: Fly ash by the US Dept. of Transportation; aerosol pollution by NASA; tree in field during extreme cold with frozen fog by Ian Furst; color differences in DOC collected from coastal waters by Epipelagic; low-temperature electron micrograph of a cluster of *E. coli* bacteria; photo by Eric Erbe; digital colorization by Christopher Pooley, both of USDA, ARS, EMU; breakdown of a plastic bottle into micro- and nanoplastics by the European Environmental Agency; microplastic fibers identified in the marine environment by M.Danny25; phospholipid aqueous solution structures by Mariana Ruiz Villarreal; MOF-5 by Tony Boehle; Fullerene-C60b by Jean-Jacques; picture of solution containing gold nanoparticles by Aleksandar Kondinski; water filtration systems by Ivan Akira; colloidal silica gel with light opalescence by Ichwarsnur; a microscopic image of a sample of human breast milk by Amangeldyurazov; Jell-O by Famartin, and a model of a DNA tetrahedron by Antony-22.

tion technology to remove contaminants. Wastewater treatment, heavy metal remediation, protection from radioactive materials, and pesticide abatement are some of the many environmental protection areas where the role of colloids is invaluable [12].

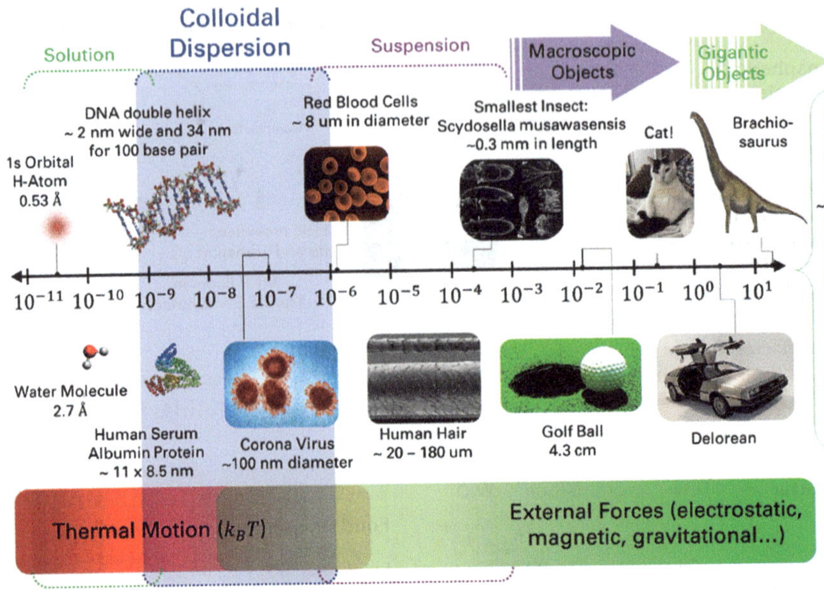

Figure 1.2: From subatomic particles to the moon and stars – size scale of matter. The length scale is in the unit of meter. Images from https://en.wikipedia.org/wiki/ via CC BY-SA 3.0, CC BY 4.0. No changes were made to the original images. Attributions: DNA animation by brian0918, ChimeraX rendering of HSA by 0fb1d8, TEM of avian infectious bronchitis virus rotated cropped Guest2625, Shape of Red Blood Cells by OpenStax College, Superfine merino wool fiber and human hair by CSIRO, *Scydosella musawasensis* by Alexey Polilov, Golfball by Lotus Head, Delorean DMC-12 side by Kevin Abato, *Brachiosaurus altithorax* side profile by Charles Nye, The Blue Marble by NASA on The Commons, The Sun by the Atmospheric Imaging Assembly of NASA's Solar Dynamics Observatory, and Milky Way Galaxy by Nick Risinger.

Application of colloids in biomedical research is a highly active area [13]. For example, dispersions of C_{60} fullerenes, which exhibit antioxidant properties, have potential application in pharmacology and in gene therapy and drug delivery. Formulated liposomes and emulsions serve as advanced modes of targeted drug delivery. They enhance bioaccessibility of hydrophobic drugs and vaccines. They are applicable in improving food digestibility and controlling the release and absorption rates of bioactive materials within the body. Given the hydrophobic/hydrophilic nature of emulsions, they are also applied for enhanced oil recovery [14]. Emulsion flooding refers to injecting a slurry of stable colloids comprising hydrocarbon, brine, and surfactants into the reservoir to mobilize the crude oil. Phase behavior and stability of emulsions are important factors in optimizing enhanced oil recovery [15]. Magnetized emulsions as sorbents for oil spills are also an attractive area of research [16, 17].

Another interesting application of colloids is related to the making of concrete – the second most consumed material on the planet after water. In general, concrete is composed of fine and coarse aggregates (e.g., sands, rocks, and gravels) and cement, mixed with water. Cement is a mixture of inorganic substances (e.g., limestones and clay) that is calcinated in a kiln up to 1450 °C and then ground into a fine powder. During the hydration process, colloidal dispersion of cement is formed. Within concrete, this cement paste acts as a binder for the aggregates and other ingredients. It is recognized that a colloidal model is needed to describe the microstructural properties of cement [18, 19]. Indeed, elucidating the colloidal properties of cement paves the way to improve strength and workability of concrete, which is literally the foundation of the modern world. Recent efforts in concrete production involve sequestering CO_2, a greenhouse gas, within the cement composites. Yet, approaches to bypass cement production altogether, because it releases extensive amounts of CO_2, are also being explored. Nevertheless, whether from cement or other substances, concrete has been in use for more than 2,000 years, dating back to the Roman architecture made from volcanic ashes from Pozzuoli, Italy.

Evidence of various other colloids being used in ancient times is also copious. For example, the ruby red solution of colloidal gold has a long history among alchemists [20]. Decoration of pottery and jewelry using colloidal gold and silver by the ancient Greeks and Romans has been noted [21]. We can also trace back to ancient Egypt to find chemical applications of colloids. For embalming purposes, colloidal clay to absorb moisture from the body was used by the Egyptians. They were utilized to remedy poisoning by the Greeks and treat various ailments by Native Americans. Egyptians also used mucilage, a natural colloid, for lubrication of large stones in building the pyramids, and for the preparation of medicines and cosmetics. Ancient civilizations, such as the Mayans, used pigments composed of minerals and plants to design pottery, ceramics, textiles, and wall paintings. Ayurvedic preparations of medicinal colloidal silver, which exhibit antimicrobial properties, can be traced back to India.

This has been only a short depiction of the use of colloids in modern times and throughout human history. Oftentimes, these colloids are concocted using natural materials, and their applications emerged rather organically. Up until relatively recently, these dispersions were utilized without the knowledge of their fundamental colloidal properties. In the modern era, we see a concerted effort for advanced applications of colloids that involve designing new and better colloids for specific purposes. This leap into the use of colloids for innovative usage required investigating and learning the basic chemistry and physics of colloids. Most importantly, recognizing that colloidal behavior is unique, that is, its attributes are different from those of bulk materials and molecules. This recognition also entails giving a name, i.e., the word “colloid,” to this interesting substance and further conglomerating this field of study into what we now call colloidal science.

We shall see more examples of colloids, along with their classifications, in the forthcoming chapter. As for what it is, it must be noted that the definition of the word “colloid” has evolved throughout time. Some key concepts, however, emerge from the aforementioned observations. Qualitatively, colloids involve tiny particles – a dispersed phase that is suspended in a continuous, bulk homogeneous medium. As we shall see in Chapter 2, the current definition of colloids involves their dimension (or size). Decades of experiments, technological advances, and theoretical developments paved the way to adequately understand colloids. It is thus only imperative that we take a brief journey through history leading to the development of colloidal science.

1.2 Colloidal Science: A Brief Historical Sketch

The history of colloidal science [22, 23] is lush and extensive. What is more, it continues to be written. This should not come as a surprise because in modern times, colloid applications are widespread. They play an important role in our society. As we have seen, colloidal chemistry is integral to the formulation of paints and coatings, food processing, pharmaceutics, nano- and biotechnology, manufacturing of personal care products, pollution remediation, aerosol science, and oil recovery. This is not an exhaustive list. Research involving colloids is vast, spanning the realm of both fundamental and applied sciences in biology, chemistry, and physics. A snapshot of the major subject areas where colloidal research is currently active is shown as a tree map in Figure 1.3. The dawn of colloidal science is closely associated with numerous experiments in the realm of physical chemistry, as such, it is the primary research area involving the fundamental properties of colloids.

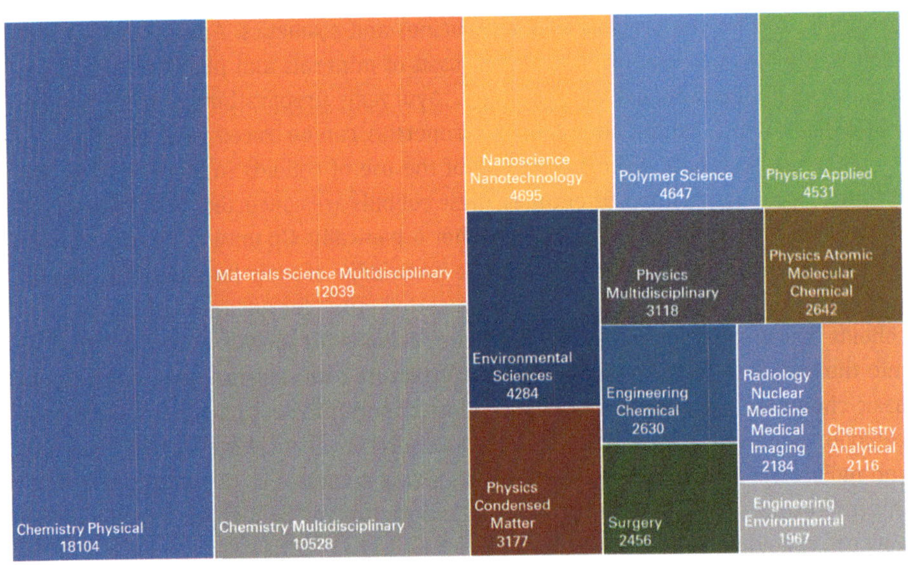

Figure 1.3: A pictorial depiction of the impact of colloidal science on different research areas – top 15 highlighted (source: Web of Science, Clarivate). The numbers represent publications related to colloids within each discipline as of October 2025.

Although the term "colloid" was not coined until 1861, interest in substances exhibiting colloidal properties can be traced back to the early 1800s. It was in 1827 that the Scottish botanist Robert Brown observed the random, zigzag-like movement of suspended particles with the aid of a microscope. The dancing granules, about 1 μm in size, were extracted from pollen of *Clarkia pulchella* [24, 25]. This incessant agitation of particles, now well recognized as Brownian motion, was not due to a biological force but was driven by thermal energy and collisions. It is the hallmark behavior of colloids that we shall discuss further. Around this time, however, colloids, as we recognize them now, did not receive a thorough scientific attention. The earliest systematic investigation of chemical systems representative of colloids appeared in 1845. The Italian chemist Francesco Selmi published his work on inorganic colloids such as silver chloride and Prussian blue and noted [26]:

> There is, however, a special type of association of solute and solvent which stands midway between solution and emulsion, because the body dwelling in the vehicle by its complete transparency would induce one to believe that it is dissolved, whereas on the contrary it is dispersed therein, in flakes, vesicles, or other shapes. [27]

Accordingly, Selmi distinguishes between "true solutions" and "pseudosolutions." The Scottish physical chemist Thomas Graham [28, 29], perhaps better known for his law of effusion, coined the term "colloid". In his 1861 paper, Graham qualitatively categorized particles into two classes – crystalloids (i.e., an aqueous solution of inorganic salts or other water-soluble molecules) and colloids (e.g., gelatin). The latter, he states, ". . . is a dynamical state of matter; the crystalloid being the statical condition" [30]. Colloids are described as substances of high molecular weight with specific molecular structure. Based on his experiment, a colloid represents substances that exhibit slower diffusion and that do not easily permeate through parchment paper.

Interestingly, based on this definition, the dispersion of finer particles as observed by Selmi and Faraday, would not be considered as colloid. Graham considered colloids to be real solutions, not pseudosolutions. Neither Selmi's work with pseudosolutions nor Michael Faraday's work on fine gold suspensions [31] was not mentioned in his paper. Whether they are called pseudosolutions, micelles, emulsions (association of liquid with liquid), demulsions (association of liquid with solid), or colloids, researchers around this time were making observations on this new state of matter that falls between the macroscopic and molecular length scales. The work of not only Selmi and Graham but also that of Baudrimont, Nageil, and Van Bemmelen contributed to this early history leading to the development of colloidal chemistry [23]. The number of scientific articles published since 1844 that contain the term "colloid" is shown in Figure 1.4. While research interest in colloids was initially slow and fluctuating, it has spiked exponentially in recent decades.

The term "colloid" stems from the Greek words *κολλα* and *ειδος* meaning "glue" and "kind," respectively. That is, the literal meaning of colloid is "glue-like." Readers will note that in our discussion of paint, ink, and cement, the term "binder" is often

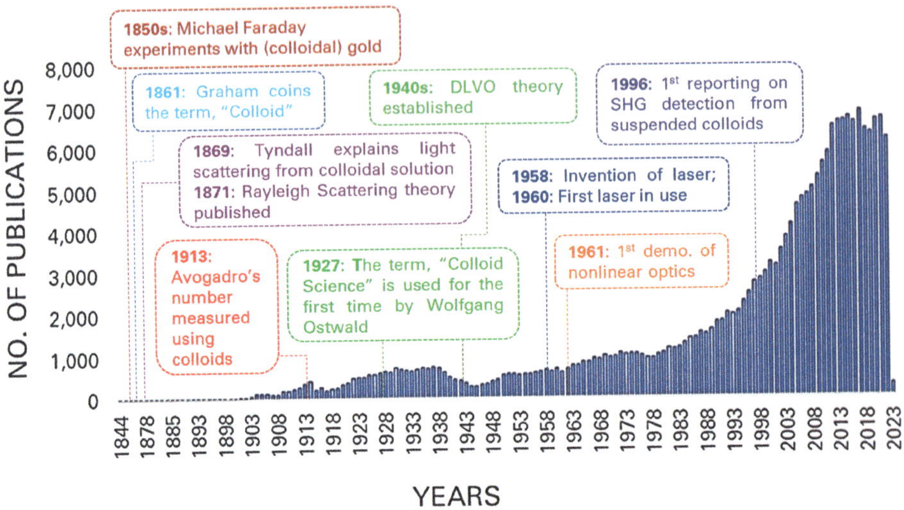

Figure 1.4: Frequency of publications containing the term "colloid" since 1844 (source: CAS SciFinder, American Chemistry Society).

used. While many colloids have glue-like properties, most do not. So, what unifies these observations and makes colloid a colloid? It is not their gluey nature, high molecular weight, or their inorganic or organic composition. Colloidal systems can be obtained with any type of matter. Up until the early 1900s, the study of colloids was in the realm of chemistry. In fact, many scientists at this time did not appreciate the special properties of colloidal systems and tried to understand them as normal solutions and by means of classical theory of homogeneous solutions. In 1927, the German chemist Wolfgang Ostwald (second son of Wilhelm Ostwald) made a note of this in his paper, the ninth edition of "The World of Neglected Dimensions" [32]. Wolfgang, along with two other scientists, Ernst A. Hauser and Aladar Buzagh, recognized that any matter, regardless of its composition, can show colloidal behavior if at least one of its dimensions falls within the colloidal range. Particle mass, number density, and compositions are not defining factors of colloid. The first experimental proof of this fact comes from the Russian chemist, P.P. von Weimarn. Based on his numerous experiments on inorganic precipitates, he concluded:

> The so-called colloidal, amorphous, and crystalloidal states are all universal (possible) properties of matter . . . it follows from these investigations that colloids and crystalloids are by no means two special worlds but that there exist close relations between themselves as well as between them and the gaseous and liquid states of matter. [22]

These research helped pinpoint the unique feature that defines colloids. It is their size or linear dimension, which had been neglected by many scientists at the time. This means colloids are pertinent not only to chemistry but also to all other disciplines. We have already seen (Section 1.1) that colloids are everywhere in nature and have vast, multifaceted applications. Wolfgang labels the pursuit of research in colloids as "colloid science," not just colloid chemistry. He founded the Kolloid Gesellschaft in Leipzig in 1922 and launched two periodicals, which brought much-needed attention to this subject matter. Buzagh, Hauser, and Wolfgang demanded further investigations on colloids. We see (Figure 1.4) a slight bump in the number of publications around this timeframe and them picking up once again after WWII. The latter is attributed to the invention of the electron microscope (1931) that allowed detailed studies of finer particles with superior resolving power.

1.2.1 Avogadro's Number

Around this time, a fundamental insight regarding colloids was emerging. It was related to the motion of the suspended particles in a dispersing medium. As was alluded to earlier, particles in the colloidal size range seem to exhibit a random chaotic motion. This movement did not originate from convection currents in the fluid or from its gradual evaporation. The motion constituted a universal phenomenon of colloidal matter. However, understanding the origin of this motion, discovered by Brown, remained elusive until the years 1888 to 1895, when the French physicist Louis G. Gouy reported his observations that the intensity of Brownian motion is directly proportional to temperature and inversely proportional to the size of the particles and the viscosity of the liquid medium. He proposed that the motion was a result of molecular and corpuscular collisions. Along came Einstein, and in 1905, he provided the theoretical foundation for Brownian motion. His theory accurately predicted the irregular, random motions of particles based on the molecular kinetic theory of heat and the premise that molecules exist [33]. He recognized that particles dispersed in a liquid exhibited osmotic pressure and that they differed from dissolved substances solely by their dimensions. Based on statistical analysis, Albert Einstein formulated an equation to calculate the mean square displacement, $\langle x^2 \rangle$, of these dancing particles. As discussed in Chapter 2, the equation relates $\langle x^2 \rangle$ to the system temperature, particle size, viscosity of the solvent, the gas constant (R), and Avogadro's number (N_A). Einstein posited that, "If the movement can actually be observed . . . the exact determination of actual atomic dimensions is then possible" [33].

Today, we take the existence of atoms or molecules for granted. It was during the 1880s the concept of atoms and molecules became scientifically prevalent. This was a tumultuous time in scientific history and whether the atoms and molecules existed was a subject of a heated scientific debate. Many prominent scientists, including Wilhelm Ostwald (one of the modern founders of the field of Physical Chemistry), refuted the notion of the atomic or molecular theory of substances. In 1909, Jean Baptiste Perrin, a French physicist, substantiated Einstein's theory and thereby the atomic theory. Perrin took the painstaking steps of preparing colloidal gamboge and mastic of uniform size and observed their trajectory with the aid of a *camera lucida*, an optical device that allows simultaneous observation and drawing of an image [34]. He measured $\langle x^2 \rangle$ of the particles and making use of the Stokes-Einstein equation (see Chapter 2), estimated the Boltzmann constant (k_B). The relationship between k_B and the gas constant allowed him to determine the Avogadro's number. He obtained a value of 7.15 $\times$ 10^{23} for N_A. This historical tidbit goes to show the role of colloids, especially their characteristic Brownian motion, in crystallizing the atomic theory and ending the dispute once and for all. Perrin's experiment, along with that of J.J. Thomson's discovery of electrons, ultimately convinced Ostwald of the existence of atoms and molecules.

The work of Einstein and Perrin established the groundwork for understanding colloids from the perspective of physics. Using gamboge, mastic, and other paint pigments, Perrin also carried out investigations on colloidal sedimentation, the process of particles settling out of liquid in the presence of a gravitational field. Guided problems at the end of this chapter explore the details of the equations (e.g., Stokes-Einstein relationship and Laplace-Perrin distribution law) involved. Qualitatively, however, it is found that the sedimentation height (z_0) is inversely proportional to the buoyant mass of the particle, which in turn is directly related to the volume of the suspended particles. The sedimentation height represents the distance of a particle above a given point of reference, and the buoyant mass represents the difference in mass density between the particle and the dispersing medium. This relationship assumes that the particle is in thermal equilibrium with the liquid molecules and that the Boltzmann distribution is applicable.

A general insight is that as the particle size or radius (r) increases, the sedimentation length decreases. As z_0 decreases, and when it is approximately equal or less than the particle size ($z_0 \lesssim r$), the particles are subject to sedimentation. They no longer remain a stable colloidal suspension. This physical insight provides a more quantitative definition of a colloid. It is a suspension of particles for which the sedimentation length is greater than or equal to the particle's linear dimension (e.g., the diameter of a spherical particle). For ordinary densities of liquids and materials, and at room temperature, the Laplace-Perrin law puts an upper limit of approximately 1 μm for the colloidal length scale.

1.2.2 The Electric Double Layer and DLVO Theory

Size is a defining factor of colloids. Another critical property that plays an important role in the overall structure of a colloidal suspension is its stability. Colloidal stability refers to the particles' ability to remain suspended in the dispersing medium. If it is unstable, its size changes by a process of aggregation or agglomeration. The stability of colloids is governed by the particle interactions with each other and the surrounding medium. In 1941, Soviet scientists Boris Derjaguin and Lev Landau offered an initial theory that explained colloidal stability. The theory is based on two opposing forces. Van der Waals forces, arising from the fluctuations in the distribution of surface electrons, provide the attraction between particles. On the other hand, electrical charges from surface-bound chemical functional groups lead to repulsion between particles. A fine balance between these forces determines whether the particles stick to each other and aggregate or stay apart and remain as a suspension. Seven years later, in 1948, two Dutch scientists, Evert Verwey and Theodoor Overbeek, independently came up with the theory and publicized it in their book. Now known as the DLVO (based on their initials) theory, it is a remarkable milestone in the history of colloidal science. This theory bypassed the shortcomings of a prior theory of Levine and Dube addressing the stability of colloidal dispersions.

The DLVO theory provides a basis for understanding colloidal aggregation and it serves as the starting point for modern theories of particle aggregation. This theory itself, however, is based on a large volume of experimental work on colloidal aggregation that preceded it since the dawn of colloidal chemistry. It also adopts various previous theoretical frameworks. For electrostatic repulsion, DLVO theory relies on a concept called the electric double layer (EDL). The EDL theory models the structure of a charged surface in contact with a liquid and the distribution of ions in its vicinity. It describes the cloud layer of ions that naturally concentrate on the charged solid surface. Tracing back to the work of Helmholtz (1850s), the major contributions to the establishment and refinement of the EDL model came from Gouy and Chapman between 1909 and 1917, and from Stern in 1924. It incorporates the Poisson-Boltzmann distribution and is an essential part of Debye-Hückel theory for the calculations of activities in electrolyte solutions. While the Gouy-Chapman model is applicable for flat surfaces, Debye-Hückel theory uses curved geometry.

We shall explore these theories in detail in Chapter 3. For now, it suffices to say that the interplay of repulsion and attraction forces that determine colloidal stability originates at the interface between the particle and its surrounding medium. The chemical compositions that terminate at the particle boundary will dictate its charge properties. The source of the van der Waals-type interactions, and additional hydrophobic and steric factors that contribute to particle stability, are also at the interfacial region between the particle and the solvent. This is yet another evidence of how investigation into colloids is ever so linked to interfacial science. Experimentally, zeta (ζ) potential, a charge characteristic property of colloids measured by light scattering, is used in DLVO theory to describe the magnitude of the electrostatic forces. Light scattering is a pivotal methodology for studying colloids, a topic we shed some light on next.

1.3 “Let There Be Light”: Measuring Colloidal Properties

Properties of colloids can be ascertained from how they interact with light. Colloids are known to scatter light. It was John Tyndall, a colleague of Michael Faraday at London’s Royal Institution, who demonstrated that a light beam passing through a colloidal suspension is visible. Recall that Faraday discovered gold sol and noted this light scattering property. This is not the case with a true solution. This phenomenon, now known as the Tyndall effect, is oftentimes employed to detect the presence of colloids. Given the colloidal dimension, the scattered visible light is intense. The observer sees the path of the light as it collides with the particles in its path within the suspension. Tyndall's work in the late 1860s also suggested that a suspension of finer particles, without absorptive properties, appears blue whereas particles of larger size appear red. Clearly, the properties of the scattered visible light detected are dependent on the colloidal size.

Various mathematical descriptions have been developed throughout history to provide a quantitative understanding of light scattering. These include the Fraunhofer, Rayleigh, and Mie scattering theories. In Chapter 4, we shall explore these theories in the context of how light is used to measure the size and distribution of particles in a colloidal suspension. Briefly, Rayleigh scattering theory, developed by Lord Rayleigh (John William Strutt) in 1871, is applicable for particles much smaller (less than 1/10th) than the wavelength of the light. Assuming red light of 632 nm, the light scattering behavior of particles less than ~60 nm can be quantitatively understood by Rayleigh's theory. This is the lower end of the colloidal size range. When particles are small compared to the wavelength of the interacting light, the intensity of the scattered light varies with the wavelength of light. As Tyndall observed, shorter wavelengths (blue light) are scattered more strongly. The Rayleigh scattering theory thus explains why the sky is blue. The blue light scattered off the small molecules in the atmosphere dominates. Mie scattering, also known as the Lorenz-Mie-Debye theory, is a rigorous mathematical solution to Maxwell's equation for light scattering by spherical particles of arbitrary materials. Published in 1908, Mie theory is applicable to colloids of a wide range of size to wavelength ratios. His theory explained the color of colloidal gold and the scattering of light from clouds. Given the complex mathematical nature, requiring large computational resources, the power of the Mie theory was not recognized until the mid-1900s. Nowadays, these theories are widely used to deduce the size of colloidal particles based on light scattering measurements (Chapter 5). The Mie theory serves as the basis for new theoretical frameworks to characterize light scattering from nonspherical particles and other special cases, such as nonlinear light scattering (Chapter 6).

Light scattering serves as a major technique for characterizing colloids and is integral to the development of colloidal science. Colloids diffract and scatter light. Many colloidal properties, such as size and charge, are discerned with the use of light. We have seen that Brownian motion was discovered using a microscope. The invention of the microscope by Hans and Zacharias Janssen, a father and son team, in the late 1600s and then the development of slit- and immersion-ultramicroscope by Siedentopf and Zsigmondy to observe fine particles, paved the way to visualize the neglected dimension. Scattered light contains information about bulk and interfacial colloidal properties. In the olden days, sunlight was used. Today, lasers are often used in major colloidal characterization tools. LASER, light amplification by stimulated emission of radiation, provides a coherent and monochromatic light source. The working principle of a laser is based on Einstein's 1917 theory of stimulated emission of light. But the practical development of the laser did not come until the 1950s. Followed by the invention of the MASER (which operates in the microwave range of the spectra), the seminal paper describing the proof of concept of the optical maser (or laser) was published by Arthur Schawlow and Charles Townes in 1958. Several others, including Gordon Gould, independently developed the idea of the laser, leading to a long patent battle. In 1960, the first working laser was built by Theodore Maiman.

Early on, lasers were not powerful enough for many practical applications. Nowadays, their commercial application is widespread – think of scanners in supermarkets, eye surgery, tattoo removal, speed check, optical data storage, and communications! Lasers are invaluable in fundamental research. Of particular interest is the use of high-intensity laser beams in the realm of nonlinear optics. At high power, light-matter interaction can lead to a nonlinear response in the material. An example of such a phenomenon is called second harmonic generation (SHG). This was first demonstrated by Peter Franken, A.E. Hill, C.W. Peters, and G. Weinreich in 1961. Imagine a laser beam of red light (632 nm). Upon interaction with matter, an output blue light at 316 nm is detected. This is SHG, also known as the frequency doubling process. Two photons of 632 nm light "combine" to generate a new photon at twice the frequency. A more general process is known as sum frequency generation (SFG). It involves two photons of arbitrary (different) frequencies and the generation of a new photon that oscillates at the sum of the two incident frequencies. Both SFG and SHG fall under the category of second-order nonlinear spectroscopy. A special selection rule makes this process interfacial specific. That is, SHG and SFG can be used to study chemical interfaces selectively. Beginning in the 1980s, SHG has been applied to study planar surfaces and interfaces. In 1996, SHG was first applied to study interfacial properties of colloids suspended in liquid. Within recent decades, both SHG and SFG have played a pivotal role in understanding the chemistry and physicochemical properties at various colloidal interfaces. Their application is the subject of Chapters 6 and 7.

This brief historical overview shows that colloidal science is an immersive, interdisciplinary science. It feeds into and adapts from different fields of science, namely Chemistry, Physics, and Biology. As with any other science, innovation in experimental techniques has shaped colloidal science. It is a growing field, and with new techniques, more insights into colloids are inevitable. Their impact on our society is limitless. In Chapter 4, along with the basic colloidal properties measured by laser light scattering, a survey of the variety of major techniques (both laser- and nonlaser-based) available is provided.

1.4 Colloidal and Interfacial Science: Fundamental Questions

Despite their omnipresence in nature and usage throughout most of human history, the discovery of colloids and pinpointing their properties (physicochemical, kinetic, electrical, magnetic, optical, interfacial, etc.) have been challenging and time consuming. Through observations and careful experimentation, in conjunction with new theories, the field of colloidal chemistry matured into a pure science. In this journey, it is apparent that colloidal science is intricately intertwined with interfacial science. Thus, the fundamental questions pertinent to colloids involve not only their mechanical and physicochemical properties, but also their interfacial characteristics. Figure 1.5 highlights some of the major attributes pertaining to colloids.

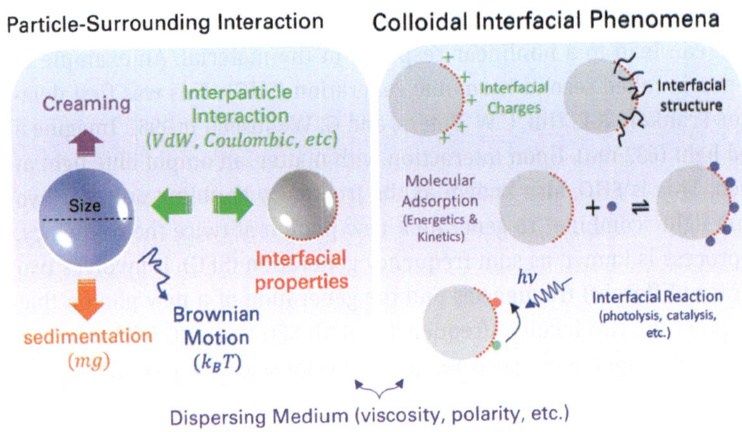

Figure 1.5: A schematic of the major fundamental properties of colloids and colloidal interfaces.

We can summarize the fundamentals of colloidal and interfacial science as follows – it involves understanding of:

(1) Particle-particle interactions and particle interaction with the dispersing medium or the solvent: These interactions depend on various forces. New types of particles are being synthesized nowadays. They can exhibit magnetic properties or asymmetry in forces as in the case of Janus particles. What is the impact of these features on colloidal stability? Recent research also focuses on the assembly of nanocolloids for the formation of thin films at various interfaces, such as liquid-liquid and air-water interfaces. Deciphering colloidal interactions is of fundamental interest.

(2) The impact of particle shape and composition: We have seen that any material can form colloids. The chemistry of these materials certainly influences colloidal behavior. Can colloids be precisely controlled, with respect to their size and shape, to optimize their applications?

(3) Interfacial properties of colloids: These include assessing their interfacial charges, polarity, and structure. In the case of semiconducting particles, elucidating surface defects is important. Colloids of soft materials are intriguing as they offer tunability of colloidal interfaces. Elucidating their interfacial structures is key for this purpose.
(4) Chemistry at colloidal interfaces: We can categorize this as molecular adsorption and reaction at the particle surface. The former includes deducing the energetics of binding and kinetics of adsorption – two key parameters important for pollution remediation technology and the fate and transport properties of contaminants. With respect to reaction, colloidal interfaces can serve to either augment or diminish reaction rates. Reaction rate parameters are important in catalysis and photodegradation pathways.
(5) Developing tools to study colloidal behavior: In particular, in situ and real time investigation of colloidal interfacial properties is a priority. Progress in science is closely linked to the invention of new techniques. The properties and processes highlighted in Figure 1.5 require experimental determination via measurements. Laser is an ideal tool to extract information from these minute systems.

This list only highlights the fundamental research in colloidal and interfacial science. However, the subject is much broader and interdisciplinary. As portrayed in Section 1.1, it includes designing new types of colloids, colloidal self-assembly, manipulating their stability, and optimizing their application. While details on the synthesis and application of colloids are beyond the scope of this book, their examples are sprinkled throughout this book, as the basic characteristics of colloids and laser applications are discussed. Chapter 7 is dedicated to a discussion of select contemporary research involving colloidal interfaces and contemporary laser-based techniques.

End of Chapter Problems

1.1 From your experience, list additional examples of colloids (not present in this chapter) found in nature and important in technology. Be sure to describe them in terms of their compositions (dispersed vs. continuous phase) and relative dimension.

1.2 Explain the significance of colloidal chemistry in (a) pollution remediation, (b) atmospheric chemistry, (c) nanotechnology, (d) drug delivery, and (e) food science.

1.3 Estimate the number of atoms contained in colloidal particles of (A) 1 μm and (B) 50 nm diameters, assuming that their densities are similar to that of water and that the average atomic mass of the atoms is 10 g/mol.

1.4 Stokes-Einstein equation for the root mean square diffusional displacement of particles undergoing Brownian Motion is $\frac{\langle x^2 \rangle}{2t} = \frac{k_B T}{6\pi\eta r}$, where t is the time, k_B is the Boltzmann constant, η is viscosity of the medium, and r is the radius of the particle. Consider the following experimental data for a particle with $r = 1 \times 10^{-6}$ m:

t (μs)	1	2	3	4	5
$\langle x^2 \rangle \times 10^{-18}$ (m^2)	0.55	0.78	1.65	1.7	2.51

(A) Determine the Boltzmann constant by plotting $\langle x^2 \rangle$ versus t and extracting the slope using a linear fit. Report standard errors.
(B) Determine the Avogadro's number, N_A, and the corresponding uncertainty.

1.5 For a spherical particle, the velocity (v) of sedimentation is given by Stokes law as follows: $v = \frac{2(\rho_2 - \rho)r^2 g}{9\eta}$, where $g = 9.8\,\frac{m}{s^2}$ is the gravitational constant. Consider particles of density (ρ_2) 1.2 kg/m^3 and of the following radii: 1 mm, 0.1 mm, 10 um, 1 um, and 10 nm. In each case, assume Stokes law to apply.
(A) Calculate the distance the particle would sediment in 1 h, in water at 20 °C (viscosity 1.002×10^{-3} kg/m s and density (ρ) 0.998 kg/m^3).
(B) Based on the data, estimate the average distance the particles traveled in 1 h due to Brownian motion. Use the Stokes-Einstein equation in problem 1.3. Note the average distance is $\sqrt{\langle x^2 \rangle}$.
(C) Comment on the distances due to sedimentation versus diffusion and the significance of colloidal dimension for the determination of Avogadro's number.

1.6 If the rate of particle sedimentation is equal to the rate of its movement due to Brownian motion, an equilibrium is reached. This is described by the Laplace-Perrin distribution law and the sedimentation height is given as $z_0 = \frac{k_B T}{m^* g}$. The buoyant mass ($m^*$) for a spherical particle depends on the particle radius (r) as follows: $m^* = \frac{4}{3}\pi \Delta\rho r^3$, where $\Delta\rho$ is the difference in mass density between the colloidal particles and the suspension medium. Determine z_0 for the particles of the following radius:

$r \times 10^{-6}$ (m)	0.005	0.01	0.05	0.1	0.5	1	10	50

Comment on the findings and the relationship between z_0 and a in regard to the definition of colloids.

1.7 Define the following terms: natural organic matter, dissolved organic matter, particulate organic matter, aerosol, DLVO theory, LASER, MASER, chemical interface (as opposed to surface), nonlinear optics, second harmonic generation, and sum frequency generation.

1.8 Discuss the possible outcomes of light-matter interaction.

1.9 Describe the influence and contributions of the following scientists to the world of colloidal and interfacial science: Robert Brown (1773–1858), Michael Faraday (1791–1867), Thomas Graham (1805–1869), Francesco Selmi (1817–1881), Lord Rayleigh (John William Strutt) (1842–1919), Richard Zsigmondy (1865–1929), Agnes Pockels (1862–1935), Jean-Baptiste Perrin (1870–1942), Albert Einstein (1879–1955), Irving Langmuir (1881–1957), Wolfgang Ostwald (1883–1943), Peter Debye (1884–1966), Theodor Svedberg (1884–1971), Katherine Blodgett (1898–1979).

2 Fundamentals of Colloids and Interfaces: Physicochemical Properties

2.1 The IUPAC Definition

According to the International Union of Pure and Applied Chemistry (IUPAC), a colloid or colloidal system is defined as “a state of subdivision, implying that the molecules or polymolecular particles dispersed in a medium have at least in one direction a dimension roughly between 1 nm and 1 μm, or that in a system, discontinuities are found at distances of that order.” This definition is quite broad. One can have a colloidal suspension with material of different dimensions. A colloidal object can be three- (3D), two- (2D), or one-dimensional (1D) as long as at least one side falls within approximately 1 nm to 1 μm. It can have arbitrary shapes and different lengths in different directions. Figure 2.1(A) depicts examples of colloids with irregular shapes within colloidal dimensions.

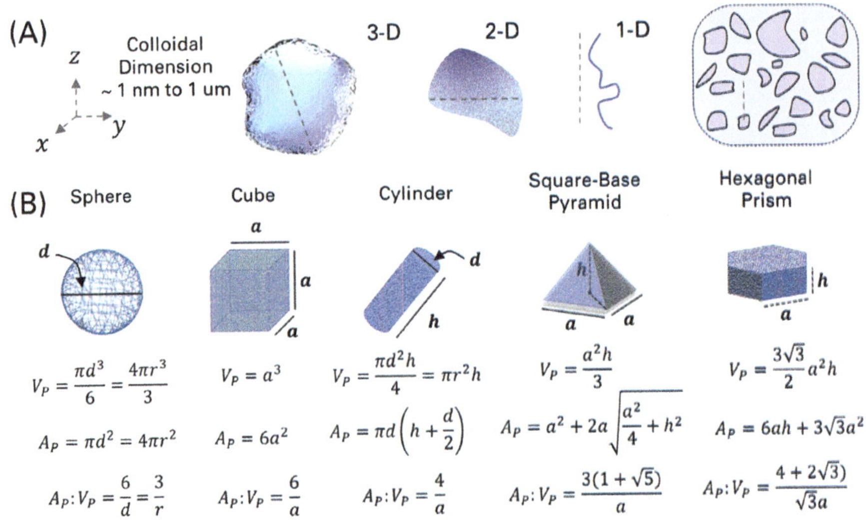

Figure 2.1: (A) Representations of different sizes and shapes of colloids. The dashed line highlights the colloidal dimension. (B) Three-dimensional geometric shapes with known volume (V_P) and surface area (A_P). In these equations, *r* and *d* represent radius and diameter, respectively. The parameter *h* is the height and *a* the length of a given side.

Both natural and man-made colloids can often be of irregular shapes. In general, a colloidal suspension can show a distribution with respect to particle size and shape. This irregularity and heterogeneity can complicate determining various physicochemical properties of colloids. Thus, fundamental questions involving colloids often in-

 | https://doi.org/10.1515/9783111018386-002

volve particles of known geometric shapes. In certain scenarios, natural colloids can be assumed to be spherical. Manufactured colloids can and do show different geometric shapes. For example, polymeric latex beads are spherical, whereas gold nanoparticles can be fabricated to be either cylindrical rods or spheres. Colloids made of semiconducting materials can form prism or hexagonal structures. Along with their equations for particle volume (V_P) and surface area (A_P), some of these shapes are illustrated in Figure 2.1(B).

As defined, the colloidal size range is a critical parameter that makes colloidal science inextricably intertwined with interfacial science. We shall demonstrate this concept using particles of known geometric shapes. Let us first consider a spherical particle. The surface area to volume ratio (A_P:V_P) for a sphere is $\frac{6}{d}$, where d represents the particle diameter. This relationship clearly shows (Figure 2.2(A)) that as the diameter of a spherical colloid decreases, the A_P:V_P ratio increases. Smaller particles have more exposed surface for the amount of space they occupy. It is especially significant for particles of submicron length scale, i.e., those within the colloidal regime. For this reason, interfacial chemistry becomes very important when colloids are considered. This relationship is not exclusive to a sphere but is true for particles of nonspherical shapes. Plots of A_P:V_P ratio versus particle volume are presented in Figure 2.2(A) for model colloids with cubic, triangular, and cylindrical geometries. The discontinuity in smoothness in nonspherical particles leads to a slight increase in the A_P:V_P ratio of these particles. Most colloids have irregular shapes; thus, assuming a spherical model would underestimate their effective surface area to a certain degree.

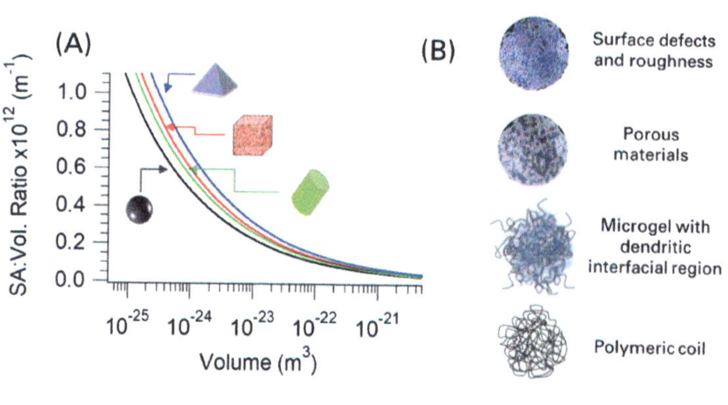

Figure 2.2: (A) A_P:V_P ratio for particles of different shapes as a function of the particle volume. The *x*-axis is on a log-scale. (B) Cartoons of three-dimensional spherical colloids with distinct surface features and porosity.

The colloidal size regime and different shapes of particles are not the only factors that dictate the interfacial chemistry of dispersed systems. The physicochemical properties of colloids and their interfacial structures play a role. Particles can have roughened surfaces or pores in them. Moreover, their surface may not be molecularly flat or smooth but can be layered, have textures, or branched structures that extend into the dispersing medium. Cartoons of such features are shown in Figure 2.2(B). The effective surface area for porous colloids and colloids with polymeric hair-like interfacial features is greater than what a smooth colloidal surface would predict. If we consider uptake of chemical species onto colloids, porous materials or branched surfaces are likely to have greater efficacy. The colloidal substance can be charged. The interface can have hydrophobic or hydrophilic characteristics. All these factors depend on the chemical composition of the colloid and the surrounding medium it is dispersed in. These attributes make colloidal science rich and extensive. Let us now explore the types of colloids that can form.

2.2 Classification of Colloids

Per IUPAC definition, the simplest picture of a colloid is that it is a two-phase system where one phase is dispersed in another. There are three common phases of matter. Accordingly, there are nine possible colloidal systems. Since gaseous species diffuse homogeneously, there are no colloids of one type of gas dispersed in another gaseous medium. Gases are miscible, and their mixture forms a true solution. The possible colloidal systems are then: solid in gas (G←S), gas in solid (G→S), liquid in gas (G←L), gas in liquid (G→L), solid in liquid (L←S), liquid in solid (L→S), liquid in liquid (L↔L), and solid in solid (S↔S). In the order of appearance, these colloidal systems are called: solid aerosol, solid foam, liquid aerosol, foam, sol or colloidal suspension (paste if solid content is high), gel, emulsion, and solid suspension (or solid dispersion). Figure 2.3 depicts these categories of colloids.

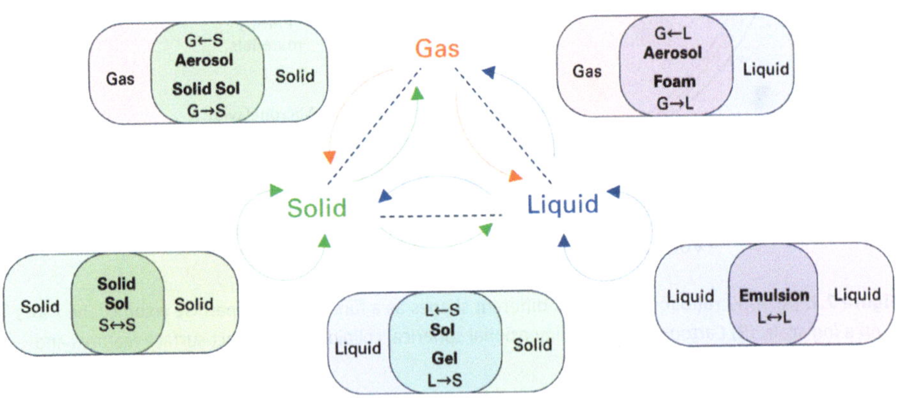

Figure 2.3: A pictorial representation of different types of colloids.

A few examples of different types of colloids are given in Table 2.1; however, it is imperative to highlight a few of these colloidal systems that are subjects of active scientific investigation. Aerosols, either G←S or G←L, are an important component of the atmosphere, and are a subject of contemporary research. There are two broad classes of aerosols – primary and secondary. Primary aerosols, such as coal fly ash, mineral dust, and sea salt, are emitted as solid particles. They can have both natural and anthropogenic sources. In contrast, secondary aerosols are formed in the atmosphere via chemical reactions under the influence of ozone, sunlight, and humidity. Major components of organic aerosols include sulfate, nitrate, and organic carbon. Colloids of gases dispersed in either solid or liquid are also ubiquitous. Polymer foam and sea foam are examples of G→S and G→L colloids, respectively. Given that water covers approximately 70% of earth's surface, finding colloids of solids in water is also common. Microplastic beads, as well as particulate natural organic matter, are solids that show colloidal behavior in liquid water. Emulsions are a prime example of L↔L colloid and these substances play important roles in the food chemistry, petroleum industry, and biomedical technologies. Targeted drug delivery entails tuning nanoemulsion interfacial chemistry. Table 2.1 shows additional examples of colloids in reference to the classifications discussed.

Table 2.1: Classification and examples of colloids.

Dispersed or particulate phase	Continuous phase or dispersing medium	Name or type of colloidal system	Examples
Gas	Liquid	Foam	Beer, froth, shaving cream, fire extinguisher, and whipped cream
Gas	Solid	Solid foam or solid sol	Marshmallow, rubber foam, lava, pumice, and Styrofoam
Liquid	Gas	Liquid aerosol	Cloud, fog, mist, and sprays
Liquid	Liquid	Emulsion	Hair cream, mayonnaise, milk, and salad dressing
Liquid	Solid	Gel	Butter, cheese, jellies, and Jell-O
Solid	Gas	Solid aerosol	Airborne bacteria, dust, fly ash, and smoke
Solid	Liquid	Sol	Cell fluids, clays, pigmented inks, paint, and toothpaste
Solid	Solid	Solid sol	Alloys, gemstones, and colored glasses

Colloids have also been categorized into two broader classes based on how the dispersed system interacts with the continuous medium, the "solvent." If the dispersed particles show attractive or favorable interactions with the continuous phase, this is known as the lyophilic (liquid-loving) colloids. The chemical functional groups present at the particle interface can allow energetically favorable interaction. Lyophilic colloids are wetted or solvated, and the interfacial interaction of solvent molecules prevents the particles from coalescing. Intrinsic colloids, which are formed readily, tend to be lyophilic. Micelles, which contain surfactants of polar headgroups and hydrophobic chains, are examples of lyophilic colloids. Protein macromolecules with various charged and polar groups also form intrinsic colloids in water. In contrast, lyophobic (liquid-fearing) colloids are not stable; their interaction with the dispersing medium is not completely energetically favorable. If the continuous phase is liquid water, the terms "hydrophilic" and "hydrophobic" are used, respectively. Perrin was the first to use the term "hydrophobic sols" in 1905 to describe gold dispersion, which exhibits a low affinity for the water.

2.3 Basic Quantitative Characteristics of Colloids

There are several physical quantities that characterize colloids. We have already seen that size and shape are important parameters. A colloidal dispersion can contain a mixture of particles with different sizes and shapes. If the size or the molecular weight of all the particles is the same, the suspension is called monodisperse or uniform. Otherwise, the suspension is a polydisperse or a nonuniform system. It is important to note that IUPAC discourages the terms "monodisperse" and "polydisperse." The term "dispersion" implies a spread. Thus, the term "monodisperse" is self-contradictory, and the latter is redundant. However, these terms are pervasive in colloidal science literature and are used in this book as well to connect to the work that already exists in this field. The term "dispersity" will be used to describe the distribution of particle properties, such as relative masses and size. Often, a size distribution exists. The size and shape of a particle determine its interfacial area. We can talk about the interfacial area of a single particle or the total interfacial area of all the particles in a given volume of the suspension. The latter requires knowledge of how many particles are in the suspension. Thus, another characterization of colloids is their concentration or number density. Herein, we summarize some of the basic colloidal properties. They are size, dispersity, molecular weight, concentration or number density, interfacial area, and average separation. These rudimentary definitions and terminologies will guide the discussions in the forthcoming chapters.

2.3.1 Average Size and Nonuniformity

Uniform colloidal systems are rare. Certain polymeric particles, such as latex beads, can be fabricated with a high degree of uniformity. However, most natural, and man-made colloids are not uniform. For a system with a high degree of dispersity, it is not possible to ascribe an exact particle size or molecular weight. An average size can be calculated. The size in this context often refers to the diameter of the particle. Theoretically, there are three common approaches to determining the average size: by the diameter itself, by its area, and based on its volume. In the first scenario, a number average, $\langle d_n \rangle$, diameter is calculated as follows:

$$\langle d_n \rangle = f_i d_i \tag{2.1}$$

The term f_i is the weighting factor describing the fraction of particles having the ith diameter (d_i). In other words, it is the number of particles (n_i) with ith diameter divided by the total number of particles:

$$f_i = \frac{n_i}{\sum_i n_i} \tag{2.2}$$

Note, $\sum_i f_i = 1$. An experimental approach for measuring the particle diameters is by visualizing them under a microscope (e.g., optical or electron). Image processing software can calculate the diameter of individual particles. A histogram is generally created, from which the average size is determined.

Like eq. (2.1), the average area, $\langle A \rangle$, is defined as $\langle A \rangle = f_i A_i$, where A_i and f_i (eq. (2.2)) are the area and the frequency of the ith particle, respectively. Generally, the average particle area can be determined from adsorption experiments (Section 3.6). If we assume that the particles are spherical, the average area is related to the area-average diameter $\langle d_a \rangle$ as follows:

$$\langle A \rangle = \pi \langle d_a \rangle^2 \tag{2.3}$$

The average area can also be described as $\langle A \rangle = \pi \sum_i f_i d_i^2$. Combining this relationship with eq. (2.3), the area average diameter, $\langle d_a \rangle$, can be expressed as shown in eq. (2.4a). In a similar fashion, for spherical particles, the volume-average diameter, $\langle d_v \rangle$, is given in eq. (2.4b):

$$\langle d_a \rangle = \sqrt{\sum_i f_i d_i^2} \tag{2.4a}$$

$$\langle d_v \rangle = \sqrt[3]{\sum_i f_i d_i^3} \tag{2.4b}$$

The average particle volume can be obtained from the density. If the colloidal system is uniform in size, the quantities $\langle d_n \rangle$, $\langle d_a \rangle$, and $\langle d_v \rangle$ are equal. For a "polydispersed" suspension, $\langle d_v \rangle > \langle d_a \rangle > \langle d_n \rangle$. The ratios $\langle d_a \rangle / \langle d_n \rangle$ and $\langle d_v \rangle / \langle d_n \rangle$ thus provide a measure of dispersity. A value greater than 1 implies a nonuniform system. The greater the value, the higher the nonuniformity of the particular property in size.

A general equation that captures these different averages is as follows:

$$d_{m,n} = \left(\frac{\sum_i f_i d_i^m}{\sum_i f_i d_i^n} \right)^{\frac{1}{m-n}} \tag{2.5}$$

The value of the powers m and n describes the relationship between the property measured (e.g., volume, area) to the object. When $m = 1$ and $n = 0$, the equation yields the most commonly used arithmetic mean or average, i.e., the number-weighted size. That is, $d_{1,0} = \langle d_n \rangle$. Likewise, $d_{2,0} = \langle d_a \rangle$ and $d_{3,0} = \langle d_v \rangle$. Other combinations of m and n values are sometimes used to define an "average" size. For instance, the Sauter mean value, $d_{3,2}$, relates to the ratio of volume to the surface area.

A more complete statistical measure of the "polydispersity" is the width of the particle size distribution profile. The number (or percentage of the total) of particles of a given diameter (or a small range of diameters) plotted against the particle diameter maps out the particle size distribution. This is known as a differential distribution, and it shows the relative number of particles at each size (diameter). In Figure 2.4(A), the red trace depicts an example of a monomodal Gaussian (or normal) differential distribution. It is monomodal because it has one peak; however, the colloidal suspension represented by this distribution is not monodispersed. As can be seen, it has a spread or a width. The horizontal line at the 50% mark (i.e., at half maximum) is called the full width at half maximum (FWHM). It describes the extent of the size variation or distribution. In the example shown, FWHM is ~110 nm; whereas, the peak occurs at ~555 nm. The FWHM is related to the standard deviation (σ), which is used to calculate the dispersity (or "polydispersity index," a term discouraged by IUPAC). The peak corresponds to the modal (most frequent) diameter. If the distribution is Gaussian, which is symmetric, the modal diameter is also the mean and the median.

Another way to show the size distribution is by plotting a cumulative particle size distribution curve (the blue trace in Figure 2.4(A)). It is the sum of the differential distributions, i.e., the integral of the differential distribution curve. A cumulative distribution expresses the percentage of the number of particles within a specific particle-size range. With this plot, the percentiles can be read directly from the y-axis. For example, if we interpolate from $y = 0.75$ (Figure 2.4(A)), we get a particle diameter of 630 nm ($d_{75} = 630$ nm). The interpretation is that 75% of the sample is smaller than 630 nm. Any number of percentile values can be determined. The d_{50} (50th percentile) value represents the median size. Additional measures of dispersity can be defined based on the percentile values. An example is a quantity called Span, which is defined as $d_{90} - d_{10}$. The relative span is Span/d_{50}. This quantity reveals how far the 10% and 90% points are apart, normalized with respect to the median.

The size distribution rarely follows a normal distribution, as shown in Figure 2.4 (A). Often, it can show a log-normal distribution profile. An example of a monomodal log-normal distribution is shown in Figure 2.4(B). It shows a peak with a spread, but it is no longer symmetrical. There is a slow drop-off or a tail on the side of the larger

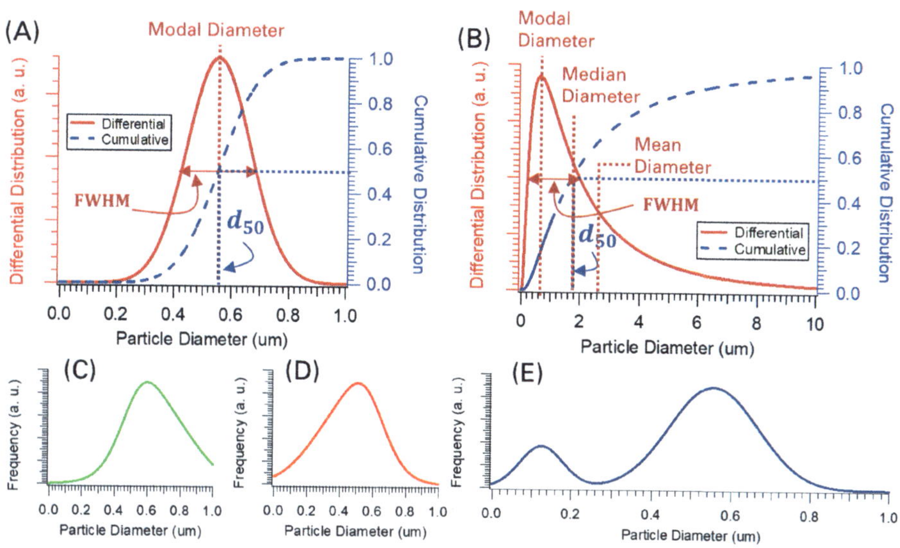

Figure 2.4: Pictorial representations of different types of size distribution, illustrating (A) normal distribution, (B) log-normal distribution, (C,D) skewed distributions, and (E) bimodal distribution.

particle size. In such scenarios, the modal, median, and the mean diameters are no longer the same. The median or d_{50} is approximately 1.8 μm in this example, whereas the modal mean is 0.64 μm. The mean diameter, which is the probability-weighted average over all possible values, is about 2.5 μm. Depending on the specific field of particle size measurements, additional terminologies can be defined to assess dispersity.

The colloidal size distribution can show skewness and multimodality. As we have seen in the log-normal plots, skewness refers to the asymmetry of the differential distribution around its modal diameter. Figure 2.4(C) and (D) portrays two distinct types of skewness – positive and negative, respectively. A positive skew represents an augmented frequency of larger-sized particles, i.e., with a “tail” toward the right (higher values). The reverse (Figure 2.4(D)) shows a negatively skewed distribution. For synthetic particles, a long tail at higher particle size is common due to diffusion-limited growth at a high degree of supersaturation. A sharper distribution appears at smaller sizes. Different modes of size distribution can exist in a colloidal sample. Multimodal distributions can be described as the sum of two or more normal (or log-normal) distributions. Figure 2.4(E) shows an example of a bimodal distribution. Two identifiable peaks are seen.

2.3.2 Average Molar Mass

Unique average values can also be defined for molar mass or molecular weight of particles in a given sample. Like eq. (2.1), the number-average molecular mass is defined as $\langle M_n \rangle = f_i M_i$, where f_i is defined by eq. (2.2). M_i is the molar mass of the ith particle. It is the ordinary arithmetic mean and is often determined by colligative methods such as osmotic pressure measurement. The mass average (or weight average) on the other hand is defined as

$$\langle M_w \rangle = \frac{\sum_i n_i M_i^2}{\sum_i n_i M_i} \tag{2.6}$$

$\langle M_w \rangle$ is often determined based on light-scattering methods. Light-scattering property depends on the size and thus the mass of the particle. As a result, larger particles contribute more compared to smaller particles. For a single-sized system, the value of $\frac{\langle M_n \rangle}{\langle M_w \rangle}$ is unity. Thus, this ratio reflects the dispersity (Đ) of the colloidal system under study. The symbol Đ (D-stroke) is introduced for dispersity to avoid confusion with the conventional use of D for the diffusion coefficient.

2.3.3 Colloidal Number Density and Volume Fraction

It is useful to describe the number of particles in a given sample as an intensive property. There are several ways to do this, and it depends on how the colloidal dispersion is prepared and the information available (i.e., particle mass, size, density, etc.). For dry particles, if the density and the volume of the particle are known, the number of particles per unit mass (n/m_p) can be determined. As an example, if the particle is spherical, $n/m = 6/\rho_p \pi d^3$, where ρ_p is the density of the particle, d is the particle diameter, and m is the total mass of the particles in a sample. This equation assumes that the particle size and density are uniform. For particles dispersed in a colloidal suspension, the particle number density ($N = n/V$) describes the number of particles per unit volume (V) of the system. While the definitions of these quantities are straightforward, determining these values can be tricky when the particles exhibit heterogeneity in size, shape, and density. In complex systems, size distributions must be considered. Clear statements on the assumptions made should be reported.

Other common quantities in describing particle concentration in a suspension are based on their volume fraction. The volume fraction of the particles in a dispersion is defined as

$$\phi \equiv \frac{nV_P}{V} = NV_P \tag{2.7}$$

Assuming density is conserved, the total density (ρ_T) of the dispersion is then

$$\rho_T = \rho_0(1-\phi) + \phi\rho_p \tag{2.8}$$

where ρ_0 is the density of the dispersant. Thus, by measuring the densities of the dispersing medium, particle, and the dispersion separately, it is possible to determine ϕ. Based on the volume fraction, the weight by volume (W_V, total mass of the particles divided by the total volume of the colloidal suspension) is shown in eq. (2.9a):

$$W_V \equiv \frac{nm_P}{V} = \phi\rho_p \tag{2.9a}$$

$$W_m \equiv \frac{nm_P}{m_T} = \frac{\phi\rho_p}{\rho_T} \tag{2.9b}$$

Likewise, particle concentration as weight by weight, mass of particles divided by the total mass of dispersion medium (m_T), is given in eq. (2.9b). Certain literature reports these values using just the volume or mass of the dispersing medium, not the total volume or mass. Thus, care must be taken in reading these values. To avoid confusion, the best practice is to be explicit when reporting these quantities.

It is important to recognize that the definition of volume fraction presented here is based on a hard-sphere model. However, no colloidal solution is composed of truly "hard" spheres. Aside from the fact that they exhibit dispersity, colloids often have softness in the interparticle potential. The elastic free energy is highest for hard spheres, and it decreases as softness increases. Thus, soft particles can experience a degree of deformation upon interaction with the dispersing medium and other particles. Soft particles or colloids generally refer to colloidal systems where the dispersed particles themselves are made of soft materials that are easily deformable by thermal fluctuations or external forces. Examples of such soft matter include polymers, gels, emulsions, foams, or biological macromolecules. The interaction potential drops sharply with distance for hard spheres, whereas it drops slowly for soft colloids. This is particularly true for liquid in liquid emulsions, microgels, and colloids of the polymer coil type [35–37]. Some of these particles are also subject to swelling, a process in which the volume of the polymer increases after absorbing liquid. Additionally, colloids are often prepared with a layer of polymeric or surface-active materials adsorbed at the particle surface. These materials exhibit softness due to their interfacial coating that serves as a stabilizer [38]. While the core of the particle may be modeled as a hard sphere, the added layer creates a thick shell, which increases the effective size (radius) of the particle. Figure 2.5(A) illustrates the difference between a hard sphere and a colloid with a hair-like polymer coating on the surface. The sum of the core radius (r_c) and shell thickness (δ) represents the effective core-shell radius (r_{cs}) in the case of the soft colloids of this type.

This altered size has an impact on the volume fraction of the particle. The effective volume fraction (ϕ_{eff}) of the particle in a dispersion is therefore

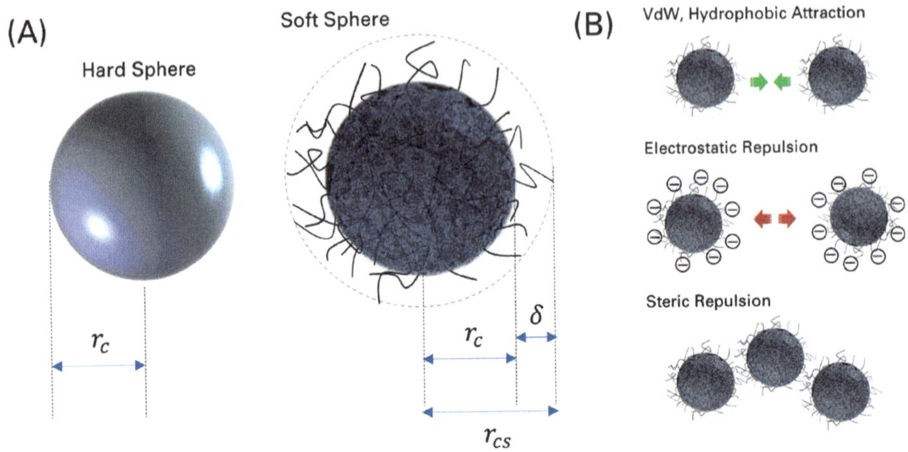

Figure 2.5: (A) Comparison of soft and hard spheres, highlighting the interfacial structure and its effect on volume fraction. (B) Different types of interparticle forces as influenced by interfacial properties of particles.

$$\phi_{\text{eff}} = \phi\left(1 + \frac{\delta}{r_c}\right) \tag{2.10}$$

The ϕ_{eff} is generally higher than the volume fraction of the core. It is apparent that the deviation from the original or "true" volume fraction (ϕ) is greatest when the particle size is small. The original volume fraction can be known from the initial volume of the dispersed phase and is typically representative of the core volume fraction. Experimentally, the volume fraction can be determined from the measurement of the particle size. This entails measuring the effective radius or diameter. Many factors, such as particle interfacial structure and charging, size dispersity, solvent effect on swelling, and porosity, influence the determination of core versus shell size. Thereby, deducing volume fraction based on size can be challenging. Another approach in determining volume fraction is based on crystallization phase behavior of particles, which also exhibits uncertainty for nonuniform particles and due to inherent assumptions.

In a recent review, it has been suggested [36] that systematic errors of $\gtrsim$3% in volume fraction are likely in practice, regardless of the technique. The key message is that the volume fraction of colloids is influenced by particle physicochemical properties. Particle volume fraction can affect particle diffusion and the viscosity of colloidal suspensions; thus, accurate measurement is necessary. The hard-sphere assumption does not always hold true. Caution must be exercised in evaluating and reporting volume fractions of particles. Colloidal softness and particle interfacial structure not only affect colloidal volume fraction but also determine its interparticle interaction and the distances between particles. Figure 2.5(B) highlights several types of interaction based on the interfacial structure of colloids. Sterically or electrostatically stabilized particles, based on interfacial bound surfactants, manifest different extents of repulsive and attractive interactions. Additionally, colloidal interfacial property can influence the total interfacial area.

2.3.4 Colloidal Interfacial Area

In Section 2.1, we have seen that the surface or interfacial area of a single particle can be calculated from the knowledge of the particle geometry. Here, we consider the total surface area pertinent to a colloidal suspension. Theoretically, the total interfacial area is $A = n A_p$. If we assume hard spheres, the total interfacial area is related to the volume fraction of the particles as follows: $A = \frac{3nV_p}{r} = \frac{3\phi V}{r}$.

The volume fraction can differ from that of the theoretically predicted for a hard sphere. Likewise, the experimental total surface area, especially for polymer-coated soft colloids and porous materials, can drastically vary from the estimates from the hard-sphere approximation. Figure 2.2 highlights the influence of particle shape on the surface area. Figure 2.6 shows various other ways the total surface area may differ from that of a smooth hard sphere. Particle surfaces can include extensions of polymer or be roughened with irregular cavities and edges. If the particle is porous, additional surface sites are available. These factors are critical, and therefore the total surface area is an important quantitative colloidal parameter.

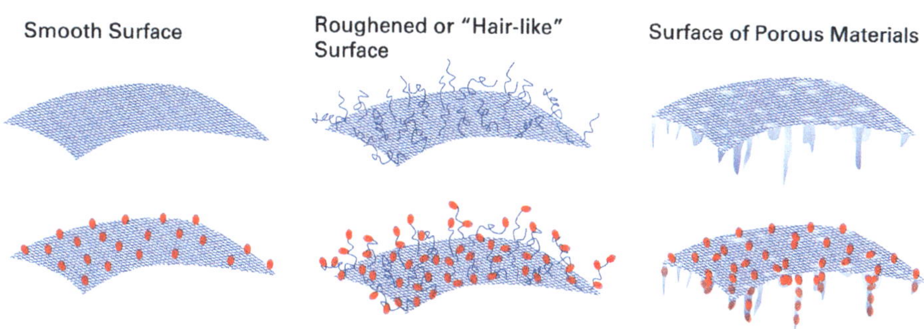

Figure 2.6: Different types of surfaces and their impact on molecular adsorption.

Experimentally, a practical definition of particle surface in use is its specific surface area (A_{SSA}). It refers to the total surface area of dry and solid particles per unit mass. A common unit is $\frac{m^2}{kg}$. Considering solid particles of spherical shape, the relationship between A_{SSA} and the particle density is then $A_{SSA} = \frac{A_P}{V_P \rho_p}$. One way of measuring this quantity is based on gas adsorption using the Brunauer-Emmett-Teller method. Brunauer-Emmett-Teller is applicable for dry particles. In a dispersion, measuring the adsorption of a given solute from solution, or its desorption from the surface, can allow measuring the specific surface area. Light-scattering techniques can yield knowledge of particle diameters (see Chapter 5). Together with particle counting, one can estimate the total surface area. However, the particle diameter based on dynamic light scattering does not reveal whether the particle is porous or if an interfacial thickness is present. Approaches that utilize adsorption mechanisms to determine particle surface area are also subject to certain challenges. For instance, the value obtained is dependent upon the energetics of adsorption for the surface. This is particularly true when adsorption from the solution phase is considered. As can be seen in Figure 2.6, multilayer formation or patch-like adsorption based on differential affinity can lead to either an over- or underestimate of the total surface area. As with volume fraction, proper reporting and adequate assessment of the interfacial area measurement are advised.

2.3.5 Average Colloidal Separation

The average separation between colloidal particles, also known as interparticle separation or spacing (IPS), depends on particle size, number density, and colloidal stability. Recall that colloids are in constant motion. Thus, the average separation represents the mean distance between the surfaces of the dispersed particles within a colloidal system. At higher concentrations, particles are closely packed together and thus exhibit smaller average separation. Repulsion between particles can cause them to stay further apart. Considering spherical particles, a simple method to approximate the average distance in a suspension is $R = \sqrt[3]{1/N}$, where N is the particle number density.

An empirical equation introduced by Funk and Dinger for calculating surface to surface distance between particles or IPS is as follows:

$$\mathrm{IPS} = \frac{2}{VSA}\left(\frac{1}{\phi} - \frac{1}{1-P_0}\right) \tag{2.11a}$$

$$\mathrm{IPS} = \frac{d}{3}\left(\frac{1}{\phi} - \frac{1}{\phi_{\mathrm{max}}}\right) \tag{2.11b}$$

where VSA (surface area to volume ratio, $A_P{:}V_P$) is the surface area per unit volume for the dry mixture of particles, ϕ is the particle volume fraction, and P_0 is the interstitial volume in the system. The term $1 - P_0$ represents the maximum volume fraction or maximum particle packing fraction, ϕ_{max}. For example, in a face-centered cubic or hexagonal close-packed arrangement, $\phi_{\mathrm{max}} \sim 0.74$. According to this equation, at the maximum loading, IPS approaches zero. When we consider a monomodal spherical particle, $\mathrm{VSA} = \frac{6}{d}$. IPS is given by eq. (2.11b). There are challenges associated with this empirical equation. Discrepancies are found when particles are porous, when the size distribution of particles is multimodal, and when the particles are in the nanoscale length scale. The latter is because the surface area-to-volume ratio increases with decreasing particle size, and the calculated IPS becomes very small (in the subnano scale), which is not physically meaningful. Even when particle size is accounted for, the Dinger-Funk model overestimates IPS at volume fractions less than ϕ_{max}. Thus, a new formula (eq. (2.12)) has been derived assuming that each particle is surrounded by a virtual cell, representing the free volume that each particle can occupy [39]:

$$\mathrm{IPS} = d\left(\left(\frac{\phi_{max}}{\phi}\right)^{1/3} - 1\right) \tag{2.12}$$

This model gives reasonable values, including cases where eq. (2.11b) yields anomalous values.

2.4 Kinetic Properties of Colloids

Colloids are in constant motion. They diffuse through the dispersing medium. Thermal, gravitational, and electrostatic forces are some of the external factors that can influence their kinetic properties. Brownian motion is heavily sensitive to temperature. Gravity or the use of a centrifuge leads to sedimentation. Charged particles are influenced by applied electric fields. Here, we review some of the basic kinetic properties of colloids. We shall note that these fundamental equations provide a connection to the measurements of various colloidal properties.

2.4.1 Translational Brownian Motion

A hallmark feature of particles in colloidal suspension is that they are in constant erratic motion. As discussed in Chapter 1, this is called the Brownian motion. It is a special case of a random walk, which is a stochastic process. Imagine that a particle can only move in one dimension, left or right, with a probability of a half for each. Let us ascribe the outcome of a flip of a coin to the direction of the particle movement – if it is a "head," the movement is one unit in the positive z-direction and if it is a "tail," the particle moves in the negative z-direction by one unit. Each time-step (say, 1 s) corresponds to the act of flipping a coin and the inherent probability is thus associated with the random nature of the direction in which the particle moves. For a 1D translational motion, a plot of the position where the particle is with respect to the initial state ($z = 0$) versus time-steps would show a jagged up and down pattern. This portrays a random walk behavior, and it involves movement that is in discrete time-steps. If the time steps are made much smaller, i.e., if the number by which we divide the time interval goes to infinity, we approach a more natural, continuous time-step. Then, the plot would describe what is observed as the Brownian movement of particles. Of course, the motion is not limited to 1D. In Figure 2.7(A), simulated Brownian

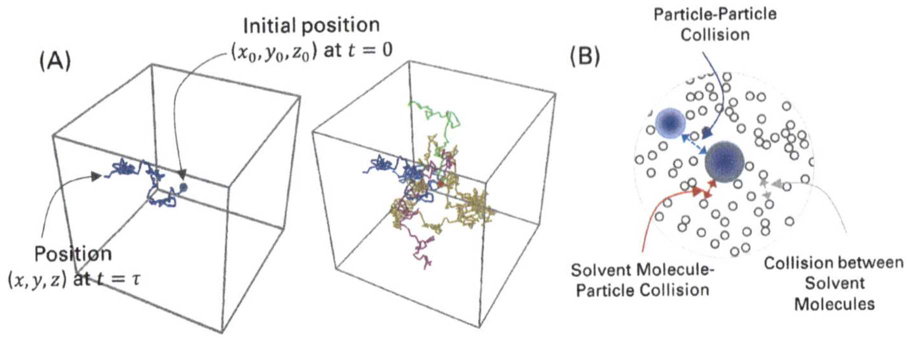

Figure 2.7: (A) Trajectory of a single particle (left) in three-dimensional space as described by Brownian motion. The right panel maps out the trajectories of four different particles starting from the origin. The simulations were generated based on the Wolfram Demonstrations Project, "Exit Times of Brownian Motion in 3D." http://demonstrations.wolfram.com/ExitTimesOfBrownianMotionIn3D/(B) Pictorial description of the different types of collisions in a colloidal suspension.

movement in 3D is shown. The left panel shows the trajectory of a single particle. The trajectories of four different particles are shown in the right panel.

The zigzag-like translational motion suggests that the particle is continuously changing its direction. The cause of this type of movement is that the particle is bombarded by tiny collisions from solvent molecules. It can also collide with yet another particle, or the wall of the container, which causes its direction to change. However, the primary source of the observed Brownian motion is that the particle is submerged in a sea of molecules that make up the bulk-dispersing medium. These solvent molecules have translational, vibrational, and rotational energies and they are under constant fluctuating motion (see Figure 2.7(B)). Under thermal fluctuation, these solvent molecules collide with the particle, and their energy is transferred, causing the particle to move in different directions. Given the randomness of the collision, the particle does not move far away from its initial position. In addition, the average value of the random variable (position in this case), is zero. In fact, if the movement is in 1D, $\langle x \rangle = 0$, i.e., the particle subject to Brownian motion is equally likely to move to the right as it is to move to the left. However, the second moment or the variance is not zero. Thus, the mean squared displacement (MSD), $\langle x^2 \rangle$, is used to represent the deviation of the position of a particle with respect to a reference position in time t. The square root of MSD, $\sqrt{\langle x^2 \rangle}$, which has the unit of length thus signifies the mean Brownian displacement or the distance of diffusion. Albert Einstein has shown that the MSD of the particle is directly proportional to time. These relationships in three different dimensions are

$$\langle x^2 \rangle = 2D_{tr}t \qquad \text{in 1D} \tag{2.13a}$$

$$\langle \rho^2 \rangle = 4D_{tr}t \qquad \text{in 2D where } \rho^2 = x^2 + y^2 \tag{2.13b}$$

$$\langle r^2 \rangle = 6D_{tr}t \qquad \text{in 3D where } r^2 = x^2 + y^2 + z^2 \tag{2.13c}$$

In these equations, the proportionality constant, *D*, is called the diffusion coefficient. It has a unit of $\frac{\text{m}^2}{\text{s}}$, and describes the amount of the solute particle that spreads out across a unit area per second. In 1D or 2D, particles will eventually return to any given small region around their starting point with certainty, a phenomenon known as recurrence. However, a particle moving in 3D has a chance of escaping indefinitely. Diffusion coefficients of some of the common colloids are given in Table 2.2. Note that the variable *r* in eq. (2.13c) represents the radial distance in the spherical coordinate, and not the radius of the particle.

The diffusion coefficient depends on a variety of factors pertinent to the particle, as well as the dispersing medium. It is directly proportional to the temperature (T) and is inversely proportional to the radius (r_p) of the particle. For a spherical particle, the translational diffusion coefficient is

$$D_{tr} = \frac{k_B T}{6\pi\eta r_p} \tag{2.14}$$

where k_B is the Boltzmann constant and η is the viscosity $\left(\frac{\text{Ns}}{\text{m}^2}\right)$ of the dispersing medium. This is known as the Einstein-Stokes equation. Demonstrated by an Irish physicist, Sir George Gabriel Stokes, the denominator in this equation represents the friction factor (f), which is related to the viscous or frictional force on a spherical body moving in a fluid. By combining eqs. (2.13c) and (2.14), we get the mean Brownian displacement:

$$\sqrt{\langle r^2 \rangle} = \sqrt{\frac{k_B T t}{\pi r_p \eta}} \tag{2.15}$$

The mean Brownian displacement increases with temperature and decreases with increasing viscosity and particle radius. This finding is of practical importance, as it shows that, by measuring the mean displacement of a particle for a given time, it is possible to deduce the particle radius. This is the basis of light scattering method for determining particle size in a dilute colloidal suspension, which we discuss in Chapter 5. At a fundamental level, as we have seen in Chapter 1, this relationship allowed the determination of Avogadro's number.

The relationship between particle size, diffusivity, and dispersant viscosity, as presented in eq. (2.14), provides an important physical insight; however, it is not always accurate in describing real colloidal systems. Aside from the fact that the particles are not always perfect spheres, there are several other factors that render this equation inadequate. For example, particles are always solvated and carry along a solvent layer as they move through the fluid medium. Particles can swell when solvated. Furthermore, at high particle density, interparticle interactions impact their movement. If the particles are charged, the interaction is significant even at low concentration. Thus, we expect the diffusion coefficient of a real colloidal suspension to deviate from that of the uniform and isotropic spherical particles in dilute solution. In practice, $f = 6\pi\eta r_p$ is the friction factor of the actual particle and $f_0 = 6\pi\eta r_{p,0}$ is defined as the friction factor of the unsolvated spherical particle with radius $r_{p,0}$. The fractional ratio $\frac{f}{f_0}$ signifies an extent of the deviation from the Einstein-Stokes equation. Factional ratios of soft, fuzzy particles, as in the case of biomacromolecules tend to be larger (Table 2.2).

Table 2.2: Physical parameters of molecules and colloidal substances in water at 20 °C.

Substance	Molecular mass (amu)	Diffusion coefficient (m^2/s)	Fractional ratio f/f_0	Sedimentation coefficient (s)
O_2 (g)	31.998	1.36×10^{-5}	–	–
H_2O (l)	18.010	25.4×10^{-10} (self-diffusion)	–	–
Sucrose	342.3	5.7×10^{-10}	–	–
Ribonuclease	13,700	1.36×10^{-10}	1.04	1.85×10^{-13}
Insulin	41,000	1.50×10^{-10}	1.2 – 1.3	$2.3 - 3.1 \times 10^{-13}$
Hemoglobin (human)	62,300	0.69×10^{-10}	1.18	4.46×10^{-13}
1 μm polystyrene bead	100,000–400,000	0.0043×10^{-10} (Calculated)	~ 1.0	$\sim 2.5 \times 10^{-8}$

2.4.2 Rotational Brownian Motion

Particles are also subject to rotational motion. Under rotational diffusion, the orientation of the particle is altered. For a spherical particle, the rotational diffusion coefficient has a similar functional form as that of translational coefficient. It is expressed as

$$D_{\text{rot}} = \frac{k_B T}{8\pi\eta r_p^3} \tag{2.16}$$

The unit of D_{rot} is s^{-1}, essentially describing the frequency of the angular motion. The mean angular displacement is $\langle\theta^2\rangle = 2D_{\text{rot}}t$. Rotational Brownian diffusion of particles plays important roles in protein docking interactions, drug binding to a targeted vessel, and in the self-assembly of colloids at an interface. The influence of particle rotation and thereby the randomized orientation is especially significant in the case of Janus particles, in which the particle surface has two or more distinct physicochemical properties.

2.4.3 Sedimentation Velocity

Brownian motion is not the only type of movement a particle experiences. Under the influence of gravity, particles can sediment. Force due to gravitation (F_g) causes the particle to move in the downward direction (see Figure 2.8(A)). A buoyant force (F_b) acts in the opposite direction. The buoyant force is the upward force on the particle exerted by the dispersing fluid. The net force ($F_g - F_b$) is directly proportional to the particle volume, V_p, and the difference between the density of the particle (ρ_p) and that of the dispersant medium (ρ_0). As the particle settles, it reaches a limiting or stationary velocity (v) due to yet another opposing force that resists its downward motion. This is called the viscous or drag force (F_v). The viscous force opposes the relative motion of the adjacent fluid layers. It is proportional to the v. The proportionality constant is called the friction factor (f) and as we have seen in Section 2.4.1, it is related to the viscosity, η, of the fluid medium. For a spherical particle, $f = 6\pi\eta r_p$.

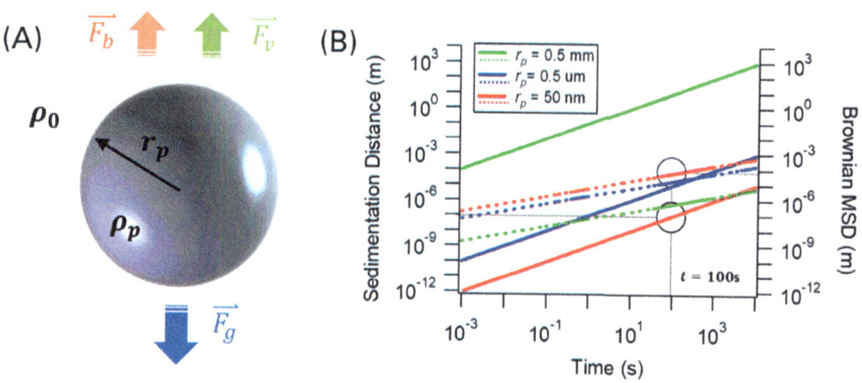

Figure 2.8: (A) Forces acting on a particle in a dispersing medium. (B) Comparison of the distance a particle travels due to sedimentation under gravity (solid traces) and Brownian motion (dotted lines). In these simulations, $\rho_0 = 0.997\frac{g}{mL}$, $\rho_p = 1.15\frac{g}{mL}$, and $\eta = 10^{-3}\frac{Ns}{m^2}$. The temperature is 293.15K and $g = 9.8\frac{m}{s^2}$.

The key idea is that the particle reaches the stationary velocity, when the net force is equal to the viscous force as follows: $V_p(\rho_p - \rho_0)g = fv$. In this relationship, g is the acceleration due to gravity. Plugging in the spherical particle volume, the corresponding friction coefficient, and rearranging the equation, an expression for the stationary-state sedimentation velocity is obtained:

$$v = \frac{2}{9}\frac{g\Delta\rho r_p^2}{\eta} \tag{2.17}$$

This is an important finding as it connects five different measurable quantities: particle size, density, velocity, dispersant density, and viscosity. Knowledge of any of these parameters allows calculation of the unknown quantity. The sedimentation velocity represents the rate at which a particle is displaced in time. In Figure 2.8(B), we see the distance particles of three different sizes sediment as a function of time. It is apparent that the larger the particle, the greater the sedimentation distance. The plot also compares estimates of the distance traveled due to sedimentation (left axis) versus Brownian motion ($\sqrt{\langle x^2 \rangle}$) (right axis). For instance, if we consider a particle of $r_p = 50$ nm at 100 s, according to eq. (2.17), it would have sedimented about 83 nm. However, the mean square displacement of the same particle upon translational diffusion (eq. (2.13c)) would cause it to travel approximately 50 um. Thus, for small particles, Brownian motion dominates and since it is in random direction, they remain suspended (assuming no aggregation) for a very long time. From Figure 2.8(B), we also see that at this timescale, the sedimentation distance and displacement due to Brownian movement become comparable for a micron-sized (the upper limit of the colloidal range) particle (blue traces). Larger particles are subject to sedimentation at a much shorter timescale.

Figure 2.8(B) highlights that the process of sedimentation can be used as a technique to separate very large particles from the small ones. To sediment smaller particles, an ultracentrifuge is necessary. In an ultracentrifuge, a colloidal suspension is rotated at speeds ranging from a few thousand revolutions per minute (RPM) to a couple of hundred thousand RPM. The result is that the particles experience a significantly enhanced gravitational pull. Modern ultracentrifuges can provide on the order of $10^6 \times g$. Under this condition, particles sediment at a much faster rate and even smaller particles succumb to sedimentation. Figure 2.9(A) shows a schematic diagram of the rotational motion and the centripetal force ($\vec{F_c}$) responsible for the faster sedimentation rate. The centripetal force results from the circular motion of an object with tangential velocity $\vec{v}$, as shown in Figure 2.9(A). The centripetal force is real and is directed inward. The term “centrifugal” force is common in this context. Centrifugal force is fictitious, and it is the apparent force that is in the opposite direction of the centripetal force. The outcome of the fast, constant, circular motion is that the suspended particles at a distance R from the center experience a movement in the same direction as the centrifugal force. These particles accumulate at the bottom of the sample holder.

The key ideas are that the particles experience a much greater acceleration than they would under a gravitational field alone and their sedimentation rate is dependent on the size of the particles. Accordingly, ultracentrifuges serve as a very useful technique to separate particles by size, as illustrated in Figure 2.9(A). Modern analytical centrifuges allow monitoring of the particle concentration gradient in situ using light absorption. Figure 2.9(B) shows a schematic diagram of this methodology. A summary of the equations pertinent to the ultracentrifuge process is as follows. The term sedimentation coefficient (s) is defined as

$$s \equiv \frac{\frac{dR}{dt}}{\omega^2 R} \tag{2.18}$$

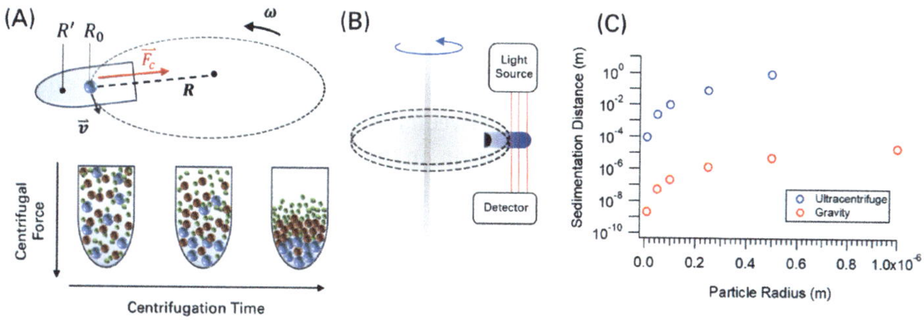

Figure 2.9: (A) Basic forces involved in an ultracentrifuge and its application in separating particles based on size. (B) A schematic diagram showing how light can be used to monitor the concentration profile of a colloidal suspension under ultracentrifugation. (C) Sedimentation distance (at $t = 60\ s$) under gravity (red markers) and centrifugal force (blue markers) as a function of particle radius. In these simulations, $\rho_0 = 0.997 \frac{g}{mL}$, $\rho_p = 1.15 \frac{g}{mL}$, and $\eta = 10^{-3} \frac{Ns}{m^2}$. The temperature is 293.15K and $g = 9.8 \frac{m}{s^2}$. The rotational speed of the centrifuge is 20,000 RPM with $R_0 = 10$ cm.

It describes the ratio of sedimentation velocity $\left(\frac{dR}{dt}\right)$ to the centrifugal acceleration. The angular velocity in the unit of radian per second is $\omega = \frac{2\pi \times \mathrm{RPM}}{60}$. The unit of s is often reported as Svedberg (S), named after the Swedish inventor of the ultracentrifuge. $1S \equiv 10^{-13} s$. Integration of eq. (2.18) yields the location of a particle along the sediment length (eq. (2.19)), where R_0 is the original position and R' describes the sedimentation distance (see Figure 2.9(A)) after time t:

$$R' = R_0 e^{\omega^2 st} \tag{2.19}$$

By measuring the radial positions R_0 and R' at different times using the absorbance method depicted in Figure 2.9(B), the sedimentation coefficient can be determined. Since s is particle density and temperature-dependent, it is measured at different concentrations at a fixed temperature and then extrapolated to zero concentration. Figure 2.9(C) compares the sedimentation distance as a function of particle radius under the influence of gravity versus ultracentrifugation.

As an example, the sedimentation coefficients are thus often reported as $s^0_{20^\circ}$, highlighting the temperature as the subscript. It is also possible to show (see Chapter End Problems) that the sedimentation coefficient is related (eq. (2.20)) to the previously discussed colloidal parameters such as the particle mass, viscosity of the dispersant, and the densities. Experimental determination of the sedimentation coefficient can allow evaluation of these and other parameters, such as diffusion coefficient and particle mass:

$$s = \frac{m}{f}\left(1 - \frac{\rho_0}{\rho_p}\right) \tag{2.20}$$

2.5 Chemical Interfaces

Interfacial science and the study of colloids are inseparable. The formation of a colloidal particle, which involves interaction between two bulk phases, is certainly affected by interfacial tension. Therefore, a discussion on the fundamental properties of a surface or interface is pertinent. First, clarification on the definition of the term "interface" is needed. An interface is a general term that describes the boundary between two distinct homogeneous phases that are in contact. The phases can be any of the bulk phases (gas, liquid, or solid). The term "surface" is specific to the case where one of the two phases is air or gas. For instance, the surface of water in contact with air is an air/water interface. In contrast, the term "surface" is not an appropriate choice in describing the boundary between oil and water in colloidal emulsions. Here, it would be called a liquid/liquid interface. Polymeric plastic microparticles suspended in water would exhibit a solid/liquid interface. The distinction between the words "surface" and "interface" is significant, especially when we consider experimental techniques used to study interfacial or surface chemistry. Many experimental tools can probe surfaces but are not selective to studying buried interfaces.

Ideally or mathematically, an interface is a 2D plane. In practice, this inhomogeneous region between two phases is not necessarily a sharp boundary but can be described as an interfacial layer (or interphase) of a finite thickness. That is, the interfacial concentration and the distribution of chemical species can vary along the perpendicular direction of the two bulk phases in contact. In fact, the interfacial layer can have a 3D thickness, normal to the plane of the interface. The dimension of this layer can be in the range of a few Angstroms (Å) to nanometers. Colloidal interfaces can also have a curvature. Nanocolloids and crystalline particles can exhibit roughness on the surface. Nevertheless, some of the basic insights about surfaces can be gained with the theoretical construction of a 2D planar interface. Herein, we consider surface properties and experimentally measurable quantities based on a planar interface model.

2.5.1 Interfacial Tension and Surface Energy

An important parameter pertaining to interfaces is associated with interfacial energetics. It relates to the work required to bring a molecule from the bulk phase to the interfacial region. Interfacial energy is associated with measurable quantities such as surface tension and surface pressure. The thermodynamics of molecular adsorption and electrostatic properties are also experimentally measurable that lend information to the concept of interfacial or surface energy. The energetics associated with an interface explain behavior such as water droplets curling into spherical balls on a hydrophobic leaf, capillary rise of liquid in a narrow space without external forces, and even why small insects such as water striders can walk on water without sinking. Figure 2.10(A) compares the curvature of a water droplet on hydrophobic Teflon and hydrophilic glass. The concepts of adhesion, cohesion, and surface tension are all related to the energy associated with the interface between two different phases. What is the source of this energy?

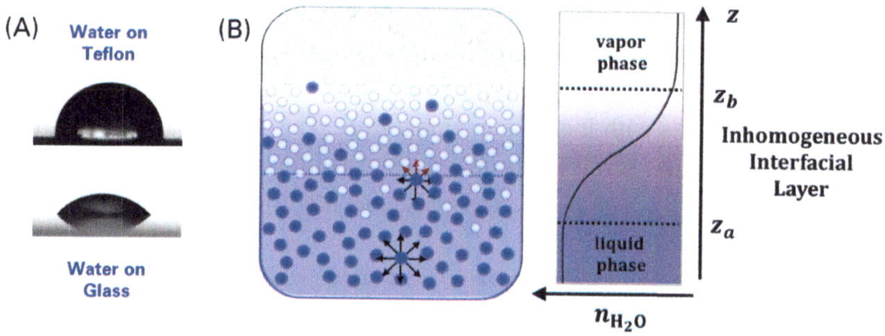

Figure 2.10: (A) Photographs of water droplets on Teflon (top) and on glass (bottom). (B) A cartoon showing the air/water interfacial region and the corresponding molecular profile.

Molecules in the interfacial region experience asymmetry in intermolecular forces. A 2D cross-section of a pure liquid phase (water) in contact with a gas (air) phase is depicted in Figure 2.10(B). The blue and white particles represent water and air molecules, respectively. The density profile of water molecules is shown to the right as a function of height (z). The interfacial layer (the region between z_a and z_b, which in general has a thickness of a few angstroms to nanometers) is a dynamic environment. Water molecules in this region can either evaporate or move into the homogeneous bulk liquid phase. Likewise, molecules in the air can diffuse into the interfacial region of the liquid. In the bulk phase, a water molecule experiences an interaction that is dominated by the surrounding water molecules. The net force is uniform, as shown by black arrows. At the interface, the same water molecule would exhibit a partial attractive force toward the water molecules below. The forces experienced due to the air phase (red arrows) are different. As a result, the interfacial region is inherently anisotropic.

Not only are the forces asymmetric, but these intermolecular forces also dictate how molecules spread or configure themselves in the interfacial region. Their affinity for each other or lack thereof determines the space they occupy. In essence, they exert a force (F) per unit length (l) along the surface. This force is related to the surface or interfacial tension, γ. In other words, the force along the line due to the interfacial bound molecules is $F = \gamma l$. A common unit of γ is mN/m; however, dyne/cm is also used. An analogy of surface tension in three-dimensions is pressure, which is defined as the tangential force exerted by molecules per unit area. Work is defined as the product of the applied force and the displacement of the object. It is the energy transferred to or from an object by a force acting on it. Accordingly, the work associated with expanding the interfacial region is therefore $w = \gamma dA$, where dA represents the infinitesimal change in the surface area. This interfacial work describes the energy required to create or increase the surface area of a material. The surface of a liquid always tends toward a minimum surface area because a system always tends to approach the state of lowest energy. Notice that the higher the surface tension of a substance is, the more work is required to expand it. Surfactants, a common ingredient in soaps, lower the surface tension of water. This is precisely why it is much easier to form bubbles with soapy water than just tap water. Another feature of surface tension is that it generally decreases with increasing temperature. This is because the intermolecular interactions that cause surface tension are disrupted due to increased molecular kinetic energy. Surface tension values of common liquids are listed in Table 2.3.

Table 2.3: Surface tension and contact angle of various liquids at 20 °C.

Liquid	Surface tension [40] in contact with air (mN/m)	Contact angle on PTFE/air θ (degrees)	Contact angle on quartz/air θ (degrees)
Mercury	486.5	150	~136
Water	72.9	112	27–50
Acetone	29.3	<10	~6
Benzene	28.9	46	~12
CCl_4	27.0	36	~9
Hexadecane	27.5	30	–
Ethanol	22.4	~38	~0

We have seen that forces between the molecules present at the interface of a liquid give rise to the physical concept of surface tension. It describes the attractive force that holds the substance together in a cohesive form. Surface tension is a physical property of a liquid in contact with a gas phase. Likewise, surface energy is the equivalent interfacial tension present between the molecules at the surface of a solid substance. Understanding surface tension and surface energy of materials is important because they are experimentally measurable quantities that tell us how strong the attraction between molecules is. These physical properties are critical to understanding how colloids form. Surface tension is key in processing various colloidal materials in fields such as medicine, drug delivery, and nanoparticle engineering. For instance, molecular adsorption at an interface alters the surface tension of a liquid. Thus, measuring the surface tension of a liquid as a function of solute concentration also serves to elucidate interfacial adsorption energy and population. Nano- or microparticles can also adsorb at the interfacial region of two liquids. In Pickering emulsions, particles attach to the oil-water interface and minimize the interfacial energy. This in turn acts as a barrier that prevents emulsion droplets from coalescing.

We conclude this section by surveying some of the approaches available to measure the surface tension of liquids. There are several methods of measuring surface tension. They include, but are not limited to, the Wilhelmy plate method, Du Noüy ring method, pendant drop method, capillary rise technique, bubble pressure method, spinning drop method, and the sessile drop method. Schematic diagrams of several of these techniques are shown. In Wilhelmy plate and Du Noüy ring methods, surface tension measurement entails pulling a liquid film with a plate or a wire ring, respectively, and measuring the force. The known geometry of the plate or the ring (see Figure 2.11) along with the measured force allows for the determination of the surface tension. The following equations relate surface tension to the measured force in the case of the Wilhelmy plate method and the Du Noüy ring method, respectively [41]:

$$\gamma = \frac{F}{2(d+l)\cos(\theta)} \tag{2.21a}$$

$$\gamma = \frac{F}{4\pi df} \tag{2.21b}$$

For the Wilhelmy plate method, a platinum plate is generally used. A complete wetting ($\theta = 0$) is often assumed. The term $2(d+l)$ represents the perimeter of the plate. For the ring method, f represents a correction factor, which is a function of the ring geometry and the liquid density. Both Wilhelmy plate and Du Noüy ring techniques require a relatively large volume of the sample.

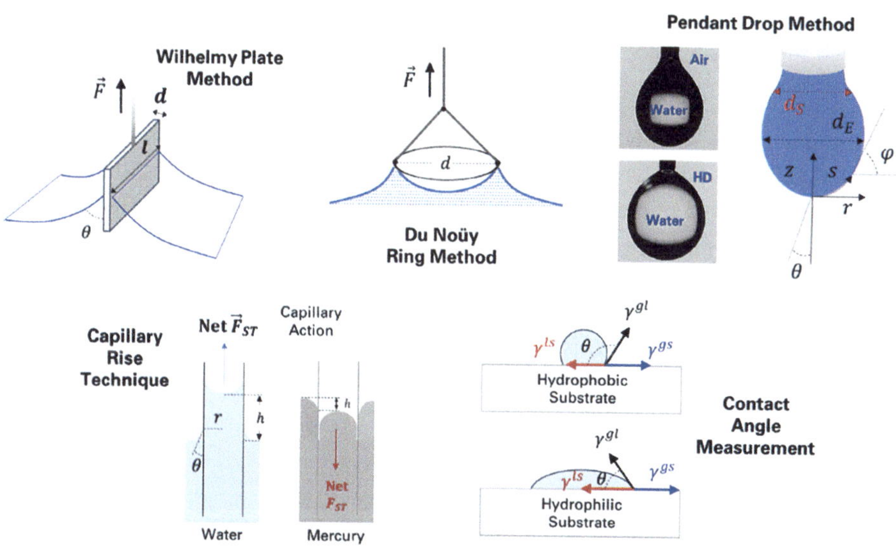

Figure 2.11: Schematic diagrams of different techniques to measure interfacial tension and contact angle of a liquid.

The pendant drop method, on the other hand, requires a small volume (~20–60 uL) of the liquid. This method involves taking a photograph of a hanging droplet and analyzing its shape. All these methods allow measurement of liquid-liquid interfacial tension. The pendant drop method in which the interfacial tensions of water droplets are measured in air and in hexadecane (HD) is shown in Figure 2.11. In these scenarios, the heavier phase is suspended within the lighter phase, such as water ($\rho_{water}^{20°C} = 0.998\,g/cm^3$) in air ($\rho_{air}^{20°C} = 1.204\,kg/m^3$) or in hexadecane ($\rho_{HD}^{20°C} = 0.773\,g/cm^3$). If the lighter phase is produced within the heavier phase, interfacial tension is measured on an ascending drop. Calculating interfacial tension using this method involves the iterative fitting of the silhouette of the pendant drop based on Young-Laplace equation, which relates the pressure difference ($\Delta\rho$) across a curved interface and the interfacial tension γ:

$$\gamma\left(\frac{1}{R_1} + \frac{1}{R_2}\right) = \Delta P_0 - \Delta\rho g z \tag{2.22}$$

In this equation, ΔP_0 is the reference pressure at $z = 0$, $\Delta\rho$ is the density difference, g is the acceleration due to gravity, R_1 and R_2 represent the principal radii of curvature. This equation is connected to a set of dimensionless differential equations relating to the arc length (s) of the droplet:

$$\frac{d\varphi}{ds} = 2 - \beta z - \frac{\sin\varphi}{r} \tag{2.23a}$$

$$\frac{dr}{ds} = \cos\varphi \tag{2.23b}$$

$$\frac{dz}{ds} = \sin\varphi \tag{2.23c}$$

The parameter $\beta = \frac{\Delta\rho g R_0^2}{\gamma}$, where R_0 is the drop radius at the apex. The Young-Laplace equation can only be solved analytically for a sphere. For other shapes, these equations must be solved numerically. This process requires computational analysis. Fitting the entire drop profile using the best theoretical curve provides greater precision for the determination of γ. A simpler estimate of the interfacial tension from a captured image of the pendant drop (Figure 2.11) is the use of eq. (2.24), where g is the acceleration due to gravity, d_e is the maximum diameter of the pendant drop, and d_s is the pendant drop diameter at a distance d_e from the bottom of the drop:

$$\gamma = \frac{\Delta\rho g d_e^2}{H} \tag{2.24}$$

The parameter H is determined empirically and is a function of the shape ratio of d_s/d_e [41].

For the capillary rise technique (Figure 2.11), surface tension is related to elevation height (h). The rise in the thin tube (capillary) is due to the surface tension at the boundary line of the liquid. The resulting force is equal to $2\pi r\gamma\cos\theta$, which is in equilibrium with the downward force associated with the weight of the liquid, $\rho g h r^2$. Equating these two forces, we get:

$$\gamma = \frac{\rho g h r}{2\cos\theta} \tag{2.25}$$

where r is the radius of the capillary and ρ is the density of the liquid. A survey of these methods clearly shows that the surface tension of a liquid can be determined in various ways.

Measurement of surface tension as a function of adsorptive concentration at a fixed temperature can be used to determine a quantity called surface or interfacial excess, Γ, based on the Gibbs adsorption isotherm equation:

$$\Gamma = -\frac{1}{RT}\left(\frac{\partial\gamma}{\partial \ln c}\right)_T \tag{2.26}$$

Equation (2.26) is applicable for an ideally dilute solution. In practice, the surface tension of a liquid at various solute concentrations is plotted. A decrease in surface tension with concentration corresponds to a positive surface excess. Surface tension can also be used to measure the critical micelle concentration (CMC). By plotting the surface tension of a solution against the logarithm of different surfactant concentrations, the CMC is determined. The concentration at which the slope of the surface tension versus logC shows a sharp leveling off, indicating that micelles are beginning to form and the surface is saturated with surfactant molecules.

Another important concept associated with surface tension is the contact angle that the liquid makes when placed on a solid substrate. The contact angle provides an indication of how well a liquid will spread over a surface (see Figure 2.11). This value is a marker of the liquid's wetting capability of a surface. Contact angle is measured via a technique called a sessile drop method. A small droplet of liquid is deposited onto a solid surface and then a camera is used to record the image. Advanced software is used to perform polynomial fitting of the droplet edge. The gradient of the polynomial is used to determine the contact angle formed at the point where the droplet touches the surface. This method is commonly used to characterize surface energies, wettability, and surface properties of materials. Referring to Figure 2.11, it is evident that when the droplet is in equilibrium, the forces are balanced and can be described by the following equation: $\gamma_{gs} = \gamma_{ls} + \gamma_{gl}\cos\theta$, where γ_{gs}, γ_{ls}, and γ_{gl} represent the interfacial tensions at the three interfaces: gas-solid, liquid-solid, and gas-liquid. Rearranging the equation gives the following, which shows the interconnection between surface tension and the contact angle θ:

$$\theta = \mathrm{Cos}^{-1}\left(\frac{\gamma_{gs} - \gamma_{ls}}{\gamma_{gl}}\right) \tag{2.27}$$

If $\gamma_{gs} > \gamma_{ls}$, then $0 < 90°$ and the droplet wets the solid. This can occur with low surface-tension liquid or high surface-energy solids such as metal. If $\gamma_{gs} < \gamma_{ls}$, then $\theta > 90°$, which represents de-wetting. This occurs with a low surface-energy solid such as hydrophobic, or a high surface-tension liquid such as water or mercury (Table 2.3). Wetting is important for the applications of coating, bonding, cleaning, painting, and printing, ensuring proper adhesion of liquids to solids. When one of the phases is solid, the wetting or adhesion energy per unit area, E_{wetting}, can be calculated using the Young-Dupré equation:

$$E_{\text{wetting}} = \gamma_{gl}(1 + \cos\theta) \tag{2.28}$$

In addition to the sessile drop method, there are other approaches to measuring contact angles. For example, contact angles can be measured as advancing (upon deposition with a needle) or receding (withdrawing with a needle). The difference between the advancing and receding contact angle is called the contact angle hysteresis. This value provides information about the surface heterogeneity and how much liquid adheres to the surface. Tilting angle or roll-off angle measurements can also be done. They are performed by tilting the stage until the droplet begins to roll off the substrate. Gravity pulls the droplet downward until it eventually rolls off the substrate, and the tilting angle is measured at the point where the drop starts moving. The leading edge of the drop is called the advancing contact angle, while the trailing edge is called the receding contact angle.

End of Chapter Problems

2.1 Define or describe the following terminologies and concepts:
- colloid
- Brownian motion
- dispersity
- interfacial tension
- emulsion
- lyophilic colloids
- lyophobic colloids
- interparticle separation
- sedimentation
- soft colloids

2.2 Given a dataset of particles with their corresponding counts and sizes:

Diameter (μm)	0.35	0.6	1.0	1.4	1.8
Number of particles	20	50	120	80	30

(A) Calculate the following particle size properties: number-average diameter (d_{10}), area-average diameter (d_{20}), volume-average diameter (d_{30}), and the Sauter mean (d_{32}).

(B) Determine the Span, relative Span, and comment on the dispersity of the sample.

2.3 Consider a spherical polystyrene nanoparticle with a radius of 100 nm suspended in water at a temperature of 25 °C.

(A) Calculate the average displacement in one dimension ($\sqrt{\langle x^2 \rangle}$) and in three dimensions ($\sqrt{\langle r^2 \rangle}$) for this particle due to translational Brownian motion.

(B) Calculate its mean angular displacement $\sqrt{\langle \theta^2 \rangle}$ due to rotational Brownian diffusion. The dynamic viscosity of water at 25 °C is 8.9×10^{-4} Ns/m^2.

2.4 A monodispersed colloidal suspension contains spherical TiO_2 particles with a diameter of 200 nm. The suspension has solids volume fraction of 0.4. Assuming a maximum packing density 0.64 (random dense packing), calculate the average interparticle spacing (IPS) between the particle surfaces using:

(A) Funk-Dinger equation

(B) Equation (2.12)

(C) Under what specific conditions does the Dinger-Funk equation for interparticle spacing provide anomalous or inaccurate values for colloidal systems, particularly when compared to equations derived from a virtual cell model or experimental data

2.5 Recall, under stationary state conditions, $V_p\left(\rho_p - \rho_0\right)g = fv$. When the force is due to centrifugal field, g can be replaced by $\omega^2 R$.

(A) Based on this information, derive eq. (2.20).

(B) Using the relationship between translational diffusion coefficient and the friction factor, show that $m = \frac{k_B T s}{D\left(1 - \rho_0/\rho_p\right)}$.

(C) The sedimentation and diffusion coefficients for myoglobin in dilute aqueous solution at 20 °C are 2.04×10^{-13}s and 1.13×10^{-10}m^2/s, respectively. The partial specific volume of the protein is 0.741 cm^3/g, the density of the solution is 1.00 g/cm^3 and the coefficient of viscosity of the solution is 1.0×10^{-3}kg/(m · s). Calculate:

i. (a) the relative molecular mass and (b) the frictional ratio of this protein.

ii. What is the probable shape of a dissolved myoglobin molecule?

2.6 Work associated with surface tension [42]:

(A) Calculate the work required to stretch a soap film such that its area increases by 5 cm^2. Note, the soap film has two sides; thus, the total increase in surface area is twice as much. The surface tension of the air-soap solution interface is 28 mN/m.

(B) A spherical drop of oil of radius 1.0 cm is dispersed as an emulsion in water. The oil molecules in the final emulsion are spherical and have a radius of 100 um. Calculate the work required for the hypothetically reversible dispersion process if the interfacial tension remains constant at 30.0 mN/m throughout.

2.7 For many binary systems, A and B, the variation of surface tension with solute concentration follows the Szyszkowski relationship: $\gamma = \gamma_A(1 - b\ln(1 + c_B/a)$, where a and b are constants and γ_A represents the surface tension of the neat solvent. Based on Gibbs adsorption isotherm equation, show that $\Gamma_B = \frac{\Gamma_B^{max} a c_B}{1 + a c_B}$, $\Gamma_B^{max} = \frac{b\gamma_A}{RT}$, and $a = 1/a$. This is known as the Szyszkowski-Langmuir isotherm.

2.8 Determination of CMC, surface excess, molecular area from surface tension measurements.

(A) Consider the following surface tension data for aqueous solutions of polyoxyethylene ($n = 4$) lauryl ether, $(C_2H_4O)_nC_{12}H_{26}O$, at 22 °C using the Wilhelmy plate method. Estimate the CMC.

Concentration (mM)	0	0.001	0.005	0.01	0.02	0.03	0.04	0.05	0.06	0.08	0.1	0.15
Surface tension (mN/m)	72.4	65.8	46.6	37.3	32.5	29.3	28.5	28.2	28.2	28.3	27.8	28.1

(B) The following surface tensions were measured for sodium dodecyl sulfate (SDS) at 22 °C using the pendant drop method. Calculate the surface excess ($mols/m^2$) and the molecular area occupied by SDS at each bulk concentration. Plot surface excess versus concentration. By fitting the data using the Langmuir model (Chapter 3), determine adsorption equilibrium constant, ΔG of adsorption, and Γ^{max}. Create a graph of molecular area versus bulk SDS concentration.

SDS concentration (mM)	0	0.5	1	1.5	2	2.5	3	3.5	4	4.5	5	5.5	6	6.5	7
Surface tension (mN/m)	72.2	66.9	64.8	61.2	56.6	52.4	51.7	50	46.7	44.4	41.9	43.1	40.2	40.6	38.9

2.9 A liquid droplet is placed on a smooth, flat, solid surface. The surface tension values for the system are as follows: $\gamma_{gl} = 72$ mN/m, $\gamma_{gs} = 60$ mN/m, and $\gamma_{ls} = 20$ mN/m.

(A) By combining eqs. (2.27) and (2.28), show that the energy of wetting or the work of adhesion can be written as $E_{\text{wetting}} = \gamma_{gl} + \gamma_{gs} - \gamma_{ls}$.

(B) Calculate E_{wetting}.

(C) Calculate the contact angle.

(D) Testing on a waterproof outdoor jacket's fabric, it is observed that a water droplet forms a contact angle of 135°. A different medical fabric, used for wound dressings, shows a contact angle of 45°. Which fabric is more hydrophilic, and what does this tell you about their intended uses?

3 Fundamentals of Colloids and Interfaces: Electrical Properties, Stability, and Adsorption

3.1 Colloidal Charge Interface

There are many ways a colloidal particle can accumulate electrostatic charges on its surface. The interfacial composition of the particle plays a major role. For instance, polymeric substances and macromolecules such as proteins have various functional groups that exhibit acid-base chemistry. Many colloidal inorganic mineral oxides (e.g., silicates, clay particles) are negatively charged. A diverse group of surfactants with inherent acid-base chemistry is also used for the preparation of emulsions. In contact with the aqueous phase, which is a high dielectric medium, these chemical groups can become protonated or deprotonated. As an example, deprotonation of carboxylic acid groups (–COOH) leads to a negatively charged interface composed of carboxylate ions, whereas protonation of amine (–NH_2) groups yields a positively charged surface. Thus, ionization serves as an exemplary process by which colloidal interfaces are charged. Certainly, the pH of the aqueous medium influences the extent of ionization and thereby the charge density at the colloidal interface.

Another mechanism by which colloidal surfaces in contact with a liquid medium acquire a net charge is through the adsorption of ions from the solution. An unequal distribution of ions at the interfacial region manifests in a net surface charge. It is observed that many colloids in natural water accumulate a negative charge on their surface. An explanation is that anions are generally more polarizable than cations and therefore less solvated by the water molecules. Cations are readily hydrated and mostly remain dissolved in the bulk [43]. Specific or preferential ion adsorption can result in charged colloids. There are also other processes that can lead to the charging of the surface. For instance, contact electrification or triboelectric charging [44–46] of particles can generate interfacial charges. Tribocharging refers to the phenomenon where charge transfer occurs as two objects come in contact or slide against each other. In such a scenario, even colloids of nonionic polymers can become charged upon contact with other surfaces. It has been suggested that water and the preferential adsorption of the hydroxide ion can explain the charging of nonionic polymeric particle surfaces.

Regardless of the mechanism, it is evident that colloids in nature often carry an interfacial charge. This has a profound implication on a variety of physicochemical processes involving colloids. Particles of like charges repel each other, keeping them far apart, thus reducing the chances of coalescence. Surface charges help to stabilize colloidal suspension. The presence of charge alters the diffusion coefficient and sedimentation rate of the particles. Charged surfaces also dictate adsorption of ions and molecules via Coulombic interaction. Therefore, understanding the charge properties of colloids is important.

 | https://doi.org/10.1515/9783111018386-003

3.2 The Electrical Double Layer: Smooth Colloids

Herein, we shall consider colloidal particles suspended in a liquid medium. The solvent molecules and dissolved ions surround the particle and create an ionic cloud around it. The physical picture describing the distribution of ionic species (Figure 3.1) and the net charge property of the colloids can be understood as an electric double layer (EDL). The concept of the EDL has undergone multiple modifications. Figure 3.1 (A) highlights the three major depictions of the charged interface throughout the late 1800s to early 1900s. Originally, the term "double layer" was coined by Hermann von Helmholtz and described the adsorption of counterions (charge with the opposite sign of that of the surface) and repulsion of coions (same charge as the surface). This combination of the oppositely charged layers forms the EDL. It can be viewed as a parallel-plate capacitor – two layers of polarized ions separated by a monolayer of solvent molecules. In this simplified picture, the EDL has approximately a single-ion thickness and within this layer, the surface potential drops linearly with distance.

The rigid nature of the double layer in this model is not a realistic representation of colloids observed in nature. This model does not consider the interactions that take place beyond the Helmholtz plane, and it does not predict the effect of electrolyte concentration on surface potential. The Gouy-Chapman (GC) model provided an improvement by applying Poisson-Boltzmann statistics to describe the distribution of ions. Developed by Georges Gouy and David Chapman, it is a more realistic model that treats the EDL as nonrigid. Ions near the charged surface are allowed to spread in the solvent due to thermal energy. The counterions are attracted to the charged interface, and the coions are repelled. They are permitted to move freely under electrostatic interaction and random thermal agitation. In this model, there is a single "diffuse" layer of ions surrounding the surface. The potential across this diffuse layer falls exponentially with distance from the surface. This model assumes that ions are point charges, that they do not interact, and that the surface is flat and uniformly charged. Moreover, the solvent is structureless; that is, the solvent medium is described by its dielectric constant.

Further updates to the EDL were made by Otto Stern, by considering the ions in the vicinity of the surface to have a defined radius, while away from the surface they act as diffused point charges. In a sense, the Stern model is a combination of the Helmholtz and the GC models. Let us explore the negatively charged colloidal surface and the ionic cloud surrounding it a bit further. Figure 3.1(B) shows a negatively charged particle dispersed in a liquid (represented as a dielectric medium) with positively and negatively charged ions surrounding the particle. This picture chiefly describes the Stern model with certain modifications made by Grahame. As we zoom in, we get a glimpse of how the positively and negatively charged ions are distributed near the surface. The counterions are attracted toward the interface, and the coions are repelled. At the very vicinity of the surface, there is a greater concentration of positively charged ions. However, due to thermal agitation and solvation, these ions

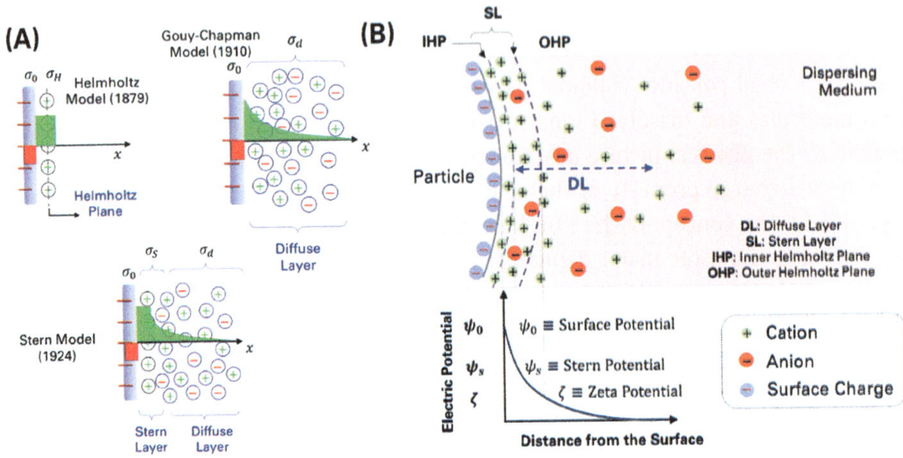

Figure 3.1: Schematic model of the electric double layer – (A) evolution of the electric double layer concept and (B) a composite picture of the Helmholtz, Gouy-Chapman, and the Stern models.

are not tightly packed at this interfacial layer. It is important to note that solvent molecules are also occupying space at the interfacial region. The ions also have a finite size, thus, the counterions at this first layer do not completely neutralize the negatively charged particle surface. Therefore, immediately away from the first layer, there is another layer of counterions. The plane that traverses the first layer of specifically adsorbed ions is called the inner Helmholtz plane. Immediately away from it, we have solvated ions, and the plane that crosses the center of the solvated ions is the outer Helmholtz plane (OHP). The region from the particle surface to the OHP is the Stern layer. It can be described as a thin compact cross section of molecules on the surface. The OHP serves as the boundary between the Stern layer and the beginning of the GC mobile diffuse layer.

Schematically, we can see that the electric potential away from the surface shows a relatively linear drop within the inner Stern layer and exponentially throughout the diffuse layer. Of importance to colloidal scientists is the zeta (ζ) potential. It is a measurable quantity that corresponds to the electric potential at which the Stern layer ends and the diffuse layer begins. It is also known as the "slipping plane" or "shear plane" – an imaginary surface of hydrodynamic shear. The shear plane is where the liquid velocity relative to the particle is zero, and it does not necessarily represent the particle surface. It is important to note that there is no direct relationship between the actual surface potential and the ζ potential. Zeta potential becomes surface potential ψ_0 if and only if the plane of shear is at the particle surface. The ζ potential can be determined based on light-scattering techniques, a topic that is discussed in Chapter 5. The EDL picture has undergone further updates, and its mathematical description is not without challenges and assumptions. Nonetheless, the basic concepts are very useful in understanding colloidal stability and charge behavior. Key summaries of this concept and its application in measuring charge properties of colloids are outlined in this section.

Based on the GC model, the potential across the diffuse layer, ψ_d, as a function of distance (x) is described by eq. (3.1), where k_B, z, e, and T represent the Boltzmann constant, valency of the counterion, the elementary charge, and the Kelvin temperature:

$$\psi_d = \frac{2k_BT}{ze}\ln\left(\frac{1+\gamma e^{-\kappa x}}{1-\gamma e^{-\kappa x}}\right) \tag{3.1}$$

The parameter γ is related to the potential at the surface, ψ_0, and is expressed as

$$\gamma = \frac{e^{\frac{ze\psi_0}{2k_BT}}-1}{e^{\frac{ze\psi_0}{2k_BT}}+1} \tag{3.1a}$$

The term κ is related to the ionic strength (I) of the electrolyte solution and is defined as

$$\kappa = \left(\frac{e^2N_A\sum_i c_iz_i^2}{\varepsilon\varepsilon_0k_BT}\right)^{\frac{1}{2}} \tag{3.1b}$$

Note, $I=\sum_i c_iz_i^2$. The parameters $\varepsilon, \varepsilon_0$, N_A, and c_i are the dielectric constant, permittivity of free space, Avogadro's number, and the concentration of the ith electrolyte species. In Figure 3.2(A,B), we see how the potential varies with distance. The positions $x=0$ is the surface and $x\to\infty$ represents the bulk solution. Figure 3.2(A) shows the effect of electrolyte concentration and its valency. Higher electrolyte concentration means larger ionic strength. Similarly, increased valency leads to an increase in ionic strength. For example, the same concentration of a divalent (2:2) electrolyte such as $MgSO_4$ has higher ionic strength than a monovalent (1:1) electrolyte, e.g., NaCl. It is observed that as ionic strength increases, there is a sharper drop in the potential across the medium. The drop mimics exponential behavior. In fact, with the Debye-Hückel approximation of $\frac{ze\psi_0}{2k_BT}\ll 1$, the surface potential can be described as $\psi_d=\psi_0e^{-\kappa x}$. This approximation is applicable at low surface potential. Figure 3.2(B) demonstrates this concept – we see a good overlap between the GC model (eq. 3.1) and the exponential depiction at $\psi_0 = 50$ mV, but a larger deviation is noted at $\psi_0 = 100$ mV. The deviation also exacerbates at low electrolyte concentration.

The inverse of the parameter κ, i.e., κ^{-1}, is known as the Debye length (λ_D). It represents a measure of the thickness of the EDL. From eq. (3.1b), we deduce that λ_D decreases with increasing ionic strength and decreasing temperature. Figure 3.1(C) shows the behavior of the Debye length as a function of electrolyte concentration. Addition of electrolyte leads to a compression of the diffuse part of the double layer. Moreover, specific adsorption of ions in the Stern layer may occur. The influence of

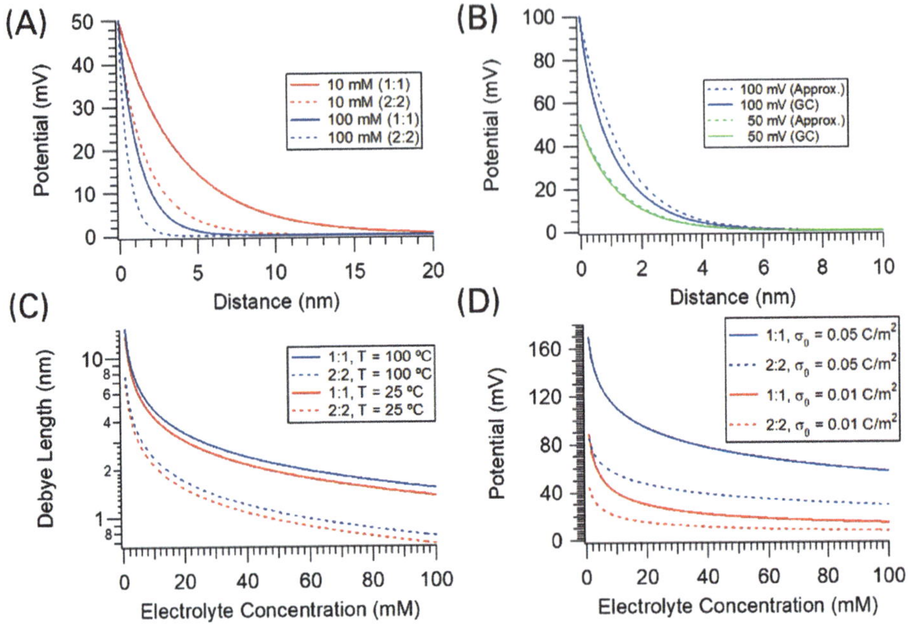

Figure 3.2: (A) The effect of electrolyte concentration and valency on the behavior of surface potential with distance as described by the GC model. In these simulations, $\psi_0 = 50$ mV. (B) Comparison of the GC model and the approximated exponential decay of the potential different ψ_0 at a fixed 1:1 electrolyte concentration of 0.1 M. (C) Debye length versus electrolyte concentration – effect of temperature and electrolyte type. (D) Relationship between potential and surface charge density as a function of electrolyte concentration. For all these simulations, unless otherwise stated, the temperature is set at 25 °C and the dielectric constant is that of water.

temperature and electrolyte type is demonstrated. In certain cases, the surface charge density is known. Grahame's equation (eq. (3.2)) provides a relationship between surface potential and the surface charge density, σ_0. At low potential, this relationship becomes $\psi_0 = \frac{\sigma_0 \lambda_D}{\varepsilon\varepsilon_0}$:

$$\psi_0 = \frac{2k_B T}{z_i e} \sinh^{-1}\left(\frac{\sigma_0}{\sqrt{8c_i \varepsilon\varepsilon_0 N_A k_B T}}\right) \tag{3.2}$$

At this juncture, it is important to highlight some of the limitations of the GC model. This theory assumes that the ions are point charges, and each ion interacts with the average electrostatic field; that is, ionic correlations are neglected. Due to the lack of a finite size, ions do not have any limit to their approach to the surface in this model. Stern's modification therefore included a thin layer of ions, now known as the Stern layer, to be specifically adsorbed at the interface. In this layer, the potential drops linearly as in the case of the Helmholtz model. The onset of the diffuse layer, typically considered as the Stern plane, is at a certain distance away from the surface. In such a scenario, ψ_0 is replaced by ψ_S in eq. (3.1). Additional assumptions are that the solvent is structure-less (described as a dielectric constant) and the surface is flat. Considering a spherical particle of radius r, eq. (3.1) is generally valid for the case where $\kappa r > 15$. As an example, for a 1 mM 1:1 electrolyte solution, the κr values of ~7 and 37 are obtained for particles of radius 100 and 500 nm, respectively. Since many colloids in nature exist within a sub-500 nm range, caution should be practiced in applying the GC model. Equation (3.3) has been shown to provide a more accurate $\sigma_0 - \psi_0$ relationship for $0.5 < \kappa r < \infty$. In addition, eqs. (3.1) and (3.2) are limited to symmetrical electrolyte of charge z. Further extensions of the GC model for asymmetrical electrolyte solutions and particles with significant curvature have been derived and are discussed elsewhere [47]:

$$\sigma_0 = \frac{2\varepsilon\varepsilon_0 k_B T}{e}\sinh\left(\frac{e\psi_0}{2k_B T}\right) \times \left[1 + \frac{1}{\kappa r}\left(\frac{2}{\cosh^2\left(\frac{e\psi_0}{4k_B T}\right)}\right) + \frac{1}{(\kappa r)^2}\left(\frac{8\ln\left[\cosh\left(\frac{e\psi_0}{4k_B T}\right)\right]}{\sinh^2\left(\frac{e\psi_0}{2k_B T}\right)}\right)\right]^{\frac{1}{2}} \tag{3.3}$$

At a broad scope, the EDL is composed of two layers – the inner compact (Stern) layer and the outer diffuse layer (as proposed by GC). A simple scenario of this inner part is that counterions, along with their hydration shells, accumulate at the interfacial region. This provides a finite thickness (δ), from the surface to the center of the counterions (i.e., OHP). In another scenario, specific adsorption of ions are tightly bound at the interface with a shorter thickness representative of the inner Helmholtz plane. Treating the Stern layer as a molecular condenser, the chief idea is that the potential difference within this region can be described as a linear function of δ as follows: $\psi_0 - \psi_d = \frac{\sigma_0 \delta}{\varepsilon\varepsilon_0}$. In this expression, ψ_d is the potential at the position where the diffuse layer begins.

3.3 Soft Particles

The EDL theory considered so far assumes a hard colloidal interface. Many types of colloids, such as colloidal natural organic matter, microplastics, biological vesicles with membranes, and various engineered nanoparticles, do not fall into the category of hard sphere. The term "soft particle" describes the three-dimensional (3D) charge distribution properties of these colloids. In these particles, the EDL is not only confined at the outskirts of the particle surface but also within its surface charge layer [48–50]. The interfacial region in contact with the liquid medium has a 3D structure and is permeable to ions from the electrolyte solution. In contrast, a hard particle does not have a surface structure or an interfacial charge layer. A hard particle is impermeable to both ions and fluid, whereas a soft particle is permeable to ions and fluid. The "soft" in this terminology is not a reference to the mechanical rigidity of the particle.

A soft particle can have a solid core, however, if the spherical colloid does not have a hard core and is a porous sphere, it is called a spherical polyelectrolyte or a porous particle. The term "polyelectrolytes" refers to macromolecules with ionizable surface functional groups covalently bonded to them. When suspended in water or other polar solvents, they accumulate charges by the dissociation of surface groups or due to adsorption of ions. Polyelectrolytes can have different types of charged groups. If both cationic and anionic groups are present, the term "polyampholyte" is used [51]. In the absence of electrolytes, these macromolecules generally adopt a linear configuration. In concentrated salt solution, where electrostatic interactions are screened, the polymers can collapse and take up a spherical or coiled configuration. Due to this extension and compression of charged polymeric chains, the surface charge properties of colloids representative of polyelectrolytes exhibit a different behavior than that of a hard particle. Examples of polyelectrolyte particles include positively charged chitosan and negatively charged DNA.

The different nature of polymers and the charged groups, as discussed in the case of polyelectrolytes, may dictate the extent of the conformational changes that occur at the interfacial region of these soft colloids. Accordingly, there is a degree of "softness" that can describe the different core and shell properties of soft particles. Five different classes of core-shell composite soft particles have been defined [50, 52]. They are illustrated in Figure 3.3. In the soft core-soft shell (Sc-Ssl) category, both the core and shell consist of soft, flexible polymers as in the case of polymer microgels. In the case of a semisoft core (SSc), e.g., natural rubber, the core is permeable to ions but not to the electroosmotic flow of the surrounding fluid.

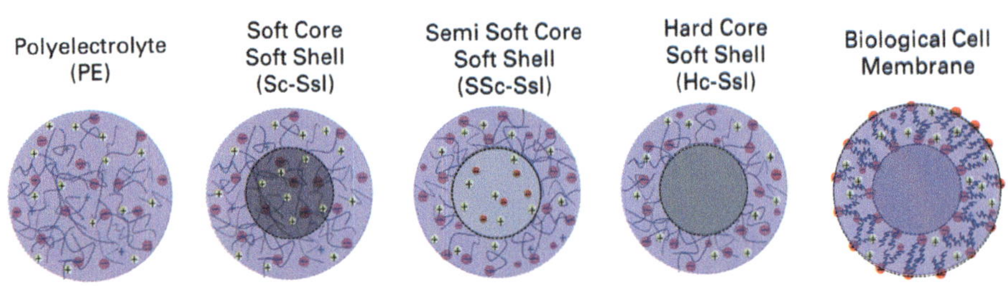

Figure 3.3: Different categories of core-shell soft particles.

As will be discussed in Section 3.4, the zeta potential of particles depends on the tangential viscosity of water, and the density of electrokinetically bound ions at the interfacial region influences this measurement. Given the nature of the soft particles, it is expected that their zeta potential behavior is drastically different than that of hard colloids. Recent advances made to tackle the challenges of understanding the charge properties of soft colloids have built upon what is already known about the hard particles. In the next section, a brief overview of the important electrostatic properties of soft colloids is introduced.

3.3.1 Surface Potential of Soft Colloids

An important feature of soft particles is that there is a potential distribution across the polyelectrolyte surface layer. The potential distribution is known as the Donnan potential, ψ_{DON}. Equation (3.4) provides a description of the Donnan potential arising due to the fixed charge density ($\rho_{fix} = ZeN$) of uniform ionized groups of valences Z. Recall that in a particle with smooth surface, surface charge density, σ_0, in the unit of $\frac{C}{m^2}$, is the only source of surface potential, ψ_0. The Donnan potential arises due to a 3D charge distribution, and the charge density in the unit of $\frac{C}{m^3}$ is used:

$$\psi_{DON} = \frac{k_B T}{ze} \sinh^{-1}\left(\frac{ZN}{2zn}\right) \tag{3.4}$$

The surface potential, ψ_0, that is, the potential at $x = 0$ (see Figure 3.4) is given as

$$\psi_0 = \psi_{DON} - \left(\frac{k_B T}{ze}\right) \tanh\left(\frac{ze\psi_{DON}}{2k_B T}\right) \tag{3.5}$$

The overall surface potential as a function of distance now involves both ψ_0 and ψ_{DON}, and is given in eq. (3.6). The parameter κ_m is the effective Debye-Hückel parameter of the polyelectrolyte layer and can be determined using eq. (3.7). Note, at low charge density, $\kappa_m = \kappa$:

$$\psi(x) = \psi_{DON} + (\psi_0 - \psi_{DON})e^{-\kappa_m |x|} \tag{3.6}$$

$$\kappa_m = \kappa\left[1 + \left(\frac{ZN}{2zn}\right)^2\right]^{\frac{1}{4}} \tag{3.7}$$

Figure 3.4 illustrates that Donnan potential diffuses not only outside but also inside the surface charge layer.

In summary, the contributions from Helmholtz, Gouy, Chapman, and Stern provide a fundamental physical understanding of these interfacial charge properties. Notably, the overall EDL is composed of diffuse and compact regions of charge distribution. In a dilute electrolyte solution, the double layer is essentially diffuse, and in a solution of high electrolyte concentration, the interface is representative of the Helmholtz description. Within the entirety of the EDL, the overall charge density is zero based on the electrical neutrality of all the charged species. An ionic cloud exists around the charged particle. We see that various factors, such as specific ion adsorption, electrolyte concentration of the solution, and temperature, can influence this thickness. Ultimately, these factors influence the stability of colloids because charge-charge interaction between particles is a major factor that governs how close two particles come into contact. In the next section, interparticle interactions and the forces responsible for colloidal stability are discussed.

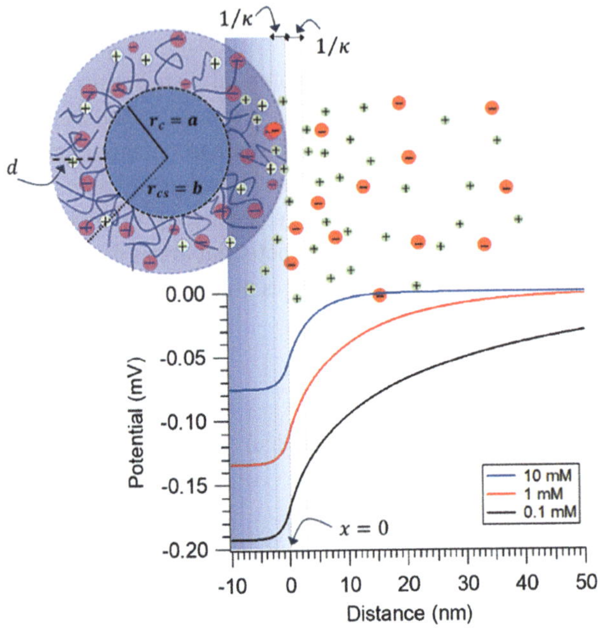

Figure 3.4: The total potential as a function of distance for a soft particle containing a polyelectrolyte layer.

It is also noted that not all colloids are hard spheres. They can exhibit soft polyelectrolyte charge layers, generating the so-called Donnan electrical potential at the solution/membrane interface. The dynamics of the electrostatic potential are then altered. The thickness and the charge density within this layer influence the overall surface potential. Current research efforts have provided insights into the electrostatic properties of soft colloids. An important measurable quantity in colloidal science is the zeta potential, which is a manifestation of the overall surface potential. Measurement of zeta potential provides an indication of colloidal stability. A particle's zeta potential is closely associated with its electrophoretic mobility (μ), i.e., the speed at which a charged colloidal particle moves in an electric field. Laser-based measurement of zeta-potential is discussed in Chapter 5. The connection between zeta potential and electrophoretic mobility is elaborated in the next section. Electrophoretic mobility behavior for soft particles, compared to that of hard smooth particles, is also highlighted.

3.4 Charged Particle Under an Electric Field

A prominent approach to gain insight into the charge properties of colloids is based on their electrokinetic properties. There are four electrokinetic effects that occur under an applied electric field in a liquid medium: electrophoresis – movement of a charged particle in a fluid; electroosmosis – movement of a bulk liquid relative to a stationary, charged surface; streaming potential – generation of an electric potential due to the flow of the liquid; and sedimentation potential – the voltage generated in an ionic or colloidal solution when charged particles move relative to the surrounding fluid. A common technique involves electrophoresis combined with light scattering from suspended particles moving under an applied electric field. Generally, the reported value based on this technique is the zeta potential. However, it is the electrophoretic mobility, μ, that is measured in this technique. ζ potential is derived from μ, and there are a lot of assumptions and caveats associated with this calculation. Thus, it is more appropriate to report μ than ζ. Electrophoretic mobility is defined as the ratio of the magnitude of the electrophoretic velocity, U, to that of the applied field, $\vec{E}$, as follows:

$$\mu = \frac{|U|}{|E|} \tag{3.8}$$

The velocity U of the particles is measured relative to that of the background electrolyte solution and is known as its rate of migration in an external electric field.

3.4.1 Electrophoretic Mobility of Hard Spherical Particle

Electrophoretic mobility is the quantity that is measured using electrophoretic light scattering (Chapter 5). It has a unit of $\frac{\mathrm{m}^2}{\mathrm{Vs}}$ (but often expressed as $\frac{\mu\mathrm{mcm}}{\mathrm{Vs}}$) and represents how easily a charged particle moves in an electric field. The larger the electrophoretic mobility, the faster the charged particles move in an electric field during electrophoresis. Highly charged particles move faster. Electrophoretic mobility is related to the surface potential ψ_0. If we assume that the slipping plane is at the particle surface, ψ_0 can be replaced by the zeta potential. The relationship can be expressed as follows:

$$\mu = \frac{\varepsilon_r \varepsilon_0}{\eta} \zeta \cdot f(\kappa r) \tag{3.9}$$

Recall (Section 3.2), zeta potential refers to the electric potential at the plane of shear relative to the potential in the bulk solution (see Figure 3.1(B)). The plane of shear, also known as the slip plane, is where the solvent molecules slip away or separate from the surface of the charged particles. As before, ε and ε_0 in eq. (3.9) are the dielectric constants, permittivity of free space, and η is the viscosity of the fluid. The parameter $f(\kappa r)$ is called Henry's function, where κ is the inverse Debye length and r represents the particle radius. The quantity is a measure of the ratio of the particle radius to the EDL thickness.

From the perspective of eq. (3.9), the limiting values of Henry's function are $\frac{2}{3}$ in the Hückel limit and 1 for the Smoluchowski case. That is, $\mu = \frac{2}{3}\frac{\varepsilon_r \varepsilon_0}{\eta} \zeta$ for the Hückel limit and $\mu = \frac{\varepsilon_r \varepsilon_0}{\eta} \zeta$ in the Smoluchowski approximation. In the Smoluchowski limit,

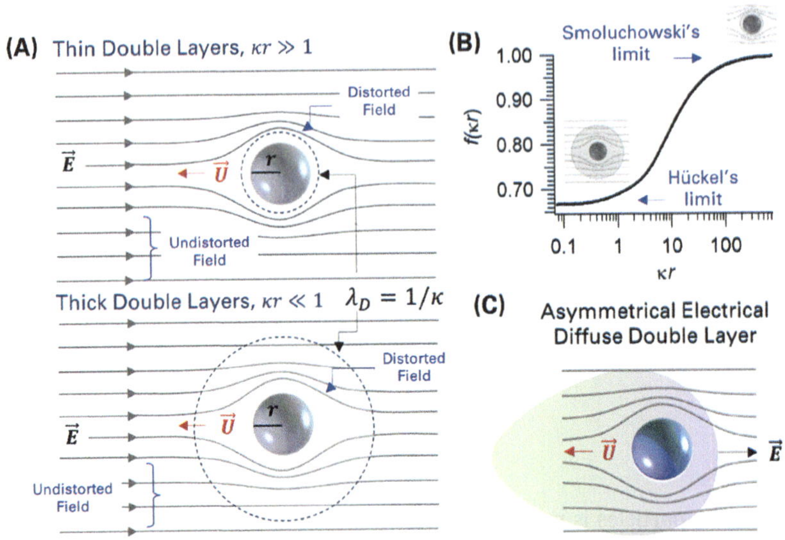

Figure 3.5: (A) Distortion of the applied electric field, $\vec{E}$, around a thin (top) versus thick (bottom) electrical double layer. The size of the dashed circle illustrates the thickness of the double layer. (B) A plot of Henry's function (eq. (3.10)). (C) Relaxation effect.

$\kappa r \gg 1$, i.e., $r \gg \lambda_D$. This means the EDL is thin relative to the particle size and the entirety of the double layer experiences the distorted field (Figure 3.5(A), top). Aqueous medium containing moderate electrolyte concentration generally satisfies the Smoluchowski limit. In the Hückel limit, $\kappa r \ll 1$, i.e., $r \ll \lambda_D$. This means the EDL is thick relative to the particle size. Most of the ions within the double layer encounter the original undistorted (parallel) field (Figure 3.5(A), bottom). Hückel approximation is generally applicable for particles in nonaqueous solvent with a low dielectric constant. Electrophoretic mobility is chiefly influenced by the ions in the EDL and near the particle surface, the distorted electric field is stronger. For this reason, the Smoluchowski mobility differ by a factor of 1.5 than the mobility obtained in the Hückel limit. Note, within certain literature, eq. (3.9) is written as $\mu = \frac{2}{3}\frac{\varepsilon_r \varepsilon_0}{\eta}\ \zeta \cdot f(\kappa r)$. In this case, $f(\kappa r)$ values are $\frac{3}{2}$ and 1 in the Smoluchowski and Hückel limits, respectively.

So, what happens in the intermediate region? For the intermediate κr values, Henry's function has a different numerical value. An approximate formula to calculate $f(\kappa r)$ is eq. (3.10). Figure 3.5(B) shows how it varies as a function of κr. Additional functional form of Henry's function for particles of spherical and other shapes are available [53]:

$$f(\kappa r) = \frac{2}{3}\left[1 + \frac{1}{2\left(1 + \frac{2.5}{\kappa r(1 + 2e^{-\kappa r})}\right)^3}\right] \tag{3.10}$$

The relationship between ζ potential and electrophoretic mobility presented so far ignores the relaxation effect – the impact of surface conductivity or double layer polarization. An excess of electrical charge at the solid/liquid interface can lead to surface conductance. It is caused by the difference in the electrical conductivity in the EDL and the bulk electrolyte solution. Surface conduction allows charges to move along the particle surface, thus can have a significant impact on the determination of the zeta potential. Generally, there is a reduction in the apparent electrophoretic mobility. However, this phenomenon is not considered in the Smoluchowski and Henry models.

The retardation of the EDL under an applied field is known as the relaxation effect. In the absence of an electric field, the ionic atmosphere is spherically distributed around the particle. The double layer is spherically symmetrical. In the presence of an applied electric field, the EDL is not necessarily symmetric. The double layer becomes distorted or polarized (Figure 3.5(C)). The net result is a drag or a backward pull on the moving charged particle. This alters the mobility of the particle. The relaxation effect is pronounced for smaller particles and with particles of high zeta potential. According to this theory [53, 54], with increased zeta potential, a surface current is generated due to the increased tangential flow of counterions in the EDL. This surface current balances the surface potential and retards the motion of the particle. A manifestation of this relaxation effect is that electrophoretic mobility becomes dependent on the ionic drag coefficient. The relationship between zeta potential and electrophoretic mobility may no longer be linear. In certain conditions, a maximum appears when the mobility of the particle is plotted against its zeta potential.

3.4.2 Electrophoretic Mobility of Soft Spherical Particle

Thus far, we have considered the electrokinetic properties of a hard sphere. For colloids with soft interfaces or polyelectrolytes (Section 3.4.1), the electrophoretic mobility not only depends on the surface potential but also on the Donnan potential and the associated Debye-Hückel parameters of the polyelectrolyte layer. An expression for the electrophoretic mobility of soft particles with relaxation effect included is eq. (3.11) [48, 55]. The relaxation effect influences the electrophoretic mobility of particles with high charge densities, such as polyelectrolytes. The parameters a and b represent the core radius (r_c) and the core-shell radius (r_{cs}) of the particle, respectively (Figure 3.4). Moreover, the expression includes ψ_{DON} in addition to ψ_0. The polyelectrolyte layer allows charges to be distributed within a volume. Thus, the charge density term ($\rho_{\mathrm{fix}} = ZeN$) appears in this equation:

$$\mu = \frac{2\varepsilon\varepsilon_0}{3\eta}\left(1+\frac{a^3}{2b^3}\right)\frac{\frac{\psi_0}{\kappa_m}+\frac{\psi_{DON}}{\lambda}}{\frac{1}{\kappa_m}+\frac{1}{\lambda}}+\frac{ZeN}{\eta\lambda^2}\left(1-\frac{F}{1+F}\cdot\frac{1}{1+e^{-\left|\frac{ze\psi_{DON}}{k_BT}\right|}}\right)$$
$$-\frac{F}{1+F}\cdot\frac{2\varepsilon\varepsilon_0}{3\eta}\left(1+\frac{a^3}{2b^3}\right)\left(\frac{k_BT}{ze}\right)$$
$$\times\left[2ln\left(\frac{1+e^{-\left|\frac{ze\psi_0}{2k_BT}\right|}}{2}\right)+\frac{\kappa}{\lambda}\left(e^{-\left|\frac{ze\psi_0}{2k_BT}\right|}-1\right)-\frac{\kappa^2}{2\lambda(\lambda+\kappa_m)}\cdot\frac{e^{\left|\frac{ze\psi_{DON}}{k_BT}\right|}-1}{1+e^{-\left|\frac{ze\psi_{DON}}{k_BT}\right|}}\right] \tag{3.11}$$

The parameter λ is related to the ratio between the frictional coefficient (γ) of the polyelectrolyte layer and the viscosity of the liquid as follows: $\lambda = \sqrt{\frac{\gamma}{\eta}}$. It characterizes the effect of fluid flow through a porous material. The quantity $\frac{1}{\lambda}$ is known as the Brinkman screening length. The relaxation effect is incorporated in the term F, which relates to the Dukhin's number, Du, associated with surface and bulk conductivities. The factor F is a dimensionless number characterizing the relaxation effect with respect to counterions, and is calculated as follows:

$$F = \frac{2}{\kappa b}(1+3m_\pm)\left(e^{\left|\frac{ze\psi_0}{2k_BT}\right|}-1\right) \tag{3.12a}$$

$$m_\pm = \frac{2}{3}\left(\frac{k_BT}{ze}\right)^2\frac{\varepsilon\varepsilon_0}{\eta D_\pm} \tag{3.12b}$$

where $m_\pm$ is the dimensionless mobility of the cation m_+ or the anion m_-, respectively. It represents the relative contribution of electroosmosis to the surface conductivity. The corresponding ionic coefficients are symbolized as $D_\pm$. Equation (3.11) is applicable when $\kappa a \gg 1, \lambda a \gg 1,\ \kappa d \gg 1,$ and $\lambda d \gg 1$. When relaxation effect is neglected (i.e., $F = 0$), eq. (3.11) reduces to

$$\mu = \frac{2\varepsilon\varepsilon_0}{3\eta}\left(1+\frac{a^3}{2b^3}\right)\frac{\frac{\psi_0}{\kappa_m}+\frac{\psi_{DON}}{\lambda}}{\frac{1}{\kappa_m}+\frac{1}{\lambda}}+\frac{ZeN}{\eta\lambda^2} \tag{3.13}$$

Figure 3.6 compares the behavior of electrophoretic mobility as a function of 1:1 electrolyte for hard particles with no relaxation effect versus soft particles with relaxation effects. A maximum in electrophoretic mobility is apparent when a relaxation effect is considered.

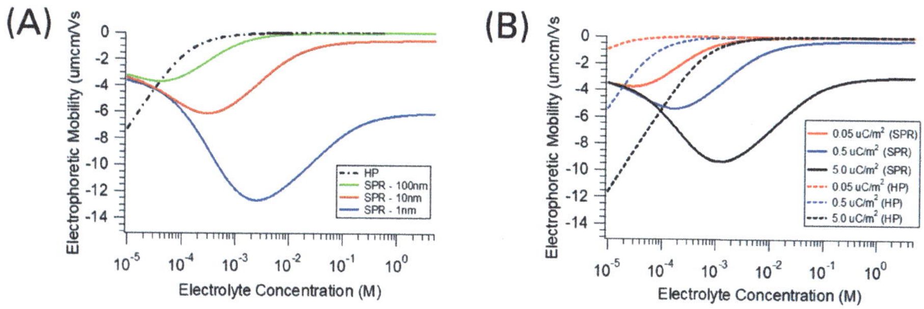

Figure 3.6: Electrophoretic mobility versus electrolyte concentrations – comparison of hard (HP) and soft particles with relaxation effect (SPR). (A) Effect of polyelectrolyte thickness ranging from 1 to 100 nm. (B) Effect of charge density. In these simulations, a = µm, $\frac{1}{\lambda}$ = 1.5 nm. The thickness of the polyelectrolyte layer for the soft particles in (B) is 10 nm.

3.5 Interparticle Interaction and Colloidal Stability

Let us start with an example in which we estimate the separation between two spherical particles. Assuming cubic packing of spheres, the average separation is the reciprocal of the cubic root of the number density. Note, in this example, we are treating the core of the particle, i.e., without an EDL or added coating on its surface. Thus, for a particle density of 10^{10}mL^{-1}, the average distance between particles is ~5 µm. The higher the number density, the smaller the separation. Particles in the colloidal size range are in constant Brownian motion and subject to sedimentation (see Figure 2.8B). The average separation for most colloidal suspensions is not large enough to overcome collisions due to their translational motion. That means, within a short time frame, these particles will eventually come in contact with each other. As they approach each other, particles can experience either attractive or repulsive forces, and the net force experienced will dictate the dispersion stability. If the net force is attractive, the particles will be subject to either agglomeration or aggregation; that is, they become colloidally unstable. The objective of this section is to outline the major sources of attractive and repulsive interaction.

Figure 2.5B highlights the three major interparticle interactions that govern colloidal stability. These interactions are (i) electrostatic repulsive interaction due to overlapping EDL, (ii) attractive interaction arising from van der Waals forces, and (iii) steric interaction. The latter is generally repulsive in nature and provides colloidal stability. The balance between the vdW and electrostatic interaction is the basis of the Derjaguin-Landau-Verwey-Overbeek (DLVO) theory of colloidal stability. A summary of this theory follows. Let $V_{\text{sph}}(R)$ describe the electrical double layer interaction free energy per unit area between two different charged hard spheres with radii r_1, r_2, and a surface separation of R. Figure 3.7 depicts the colloidal parameters for both hard and soft spheres.

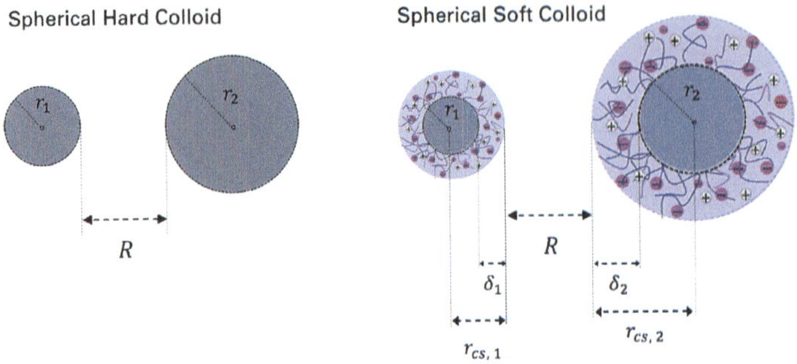

Figure 3.7: Interparticle interaction parameters.

(i) Electrostatic repulsion – for two unique hard particles, assuming $\kappa r_1 \gg 1$, $\kappa r_2 \gg 1$, $R \ll r_1$, and $R \ll r_2$, and that the surface potentials of each particle are constant, the EDL interaction energy is given as follows:

$$V_{\text{sph},R}(R) = \frac{64\pi c_i k_B T \gamma_1 \gamma_2}{\kappa^2}\left(\frac{r_1 r_2}{r_1 + r_2}\right) e^{-\kappa R} \tag{3.14}$$

This equation also assumes that the interactions between the two spheres vanish rapidly with distance and the contributions to the total interaction energy from the particles that are very far from the point of closest approach are insignificant. These assumptions are known as the Derjaguin approximation. If the particles have the same radius, r, eq. (3.14) reduces to $V_{\text{sph},\,R}(R) = \frac{32\pi c_i k_B T \gamma^2}{\kappa^2} r e^{-\kappa R}$. Expressions for γ and κ are given in eq. (3.1a) and (3.1b), respectively. For like-charged particles, this interaction is repulsive and $V_{\text{sph},R}(R)$ is positive. For two arbitrarily charged soft particles, the potential energy can be described as

$$V_{\text{sph},R}(R) = 4\pi\, \varepsilon\varepsilon_0 \left(\frac{r_{cs,1} r_{cs,2}}{r_{cs,1} + r_{cs,2}}\right) \psi_{\text{eff},1}\psi_{\text{eff},2} e^{-\kappa R} \tag{3.15}$$

where $\psi_{\text{eff},i} = \frac{4k_B T}{ze}\gamma_i$. Note, $r_{cs,i} = r_i + \delta_i$. The surface potentials are related to the Donnan potentials of the thick surface layers of the particles (Section 3.3.1).

(ii) van der Waals attraction – for two distinct spherical particles of radii r_1 and r_2, the attractive potential energy is

$$V_{\text{sph, vdW}}(R) = -\frac{A}{6}\left[\frac{y}{x^2 + xy + x} + \frac{y}{x^2 + xy + x + y} + 2\ln\left(\frac{x^2 + xy + x}{x^2 + xy + x + y}\right)\right] \tag{3.16}$$

where $x = \frac{R}{r_1 + r_2}$ and $y = \frac{r_1}{r_2}$. If the separation between the particles is small, i.e., R is much smaller than the particle radii, $V_{\text{sph, vdW}}(R)$ reduce to $\frac{Ar_p}{12R}$ for equal size particles ($r_1 = r_2 = r_p$). In eq. (3.16), the term A is called the Hamaker constant. Depending on the application, it is evaluated under conditions of either vacuum or a dielectric medium. For the latter case, an effective Hamaker constant is used. The Hamaker constant is related to the dispersion, polarization, and induction forces, and thus estimated based on atomic polarizabilities and densities (ρ) of the materials involved. When the medium is vacuum, the Hamaker constant can be calculated as $A = \pi^2 C \rho^2$, where C is the London coefficient [56, 57]. However, when working with colloidal particles in water medium, a more accurate formula based on the Lifshitz approach is used. Accordingly, the Hamaker constant of particles 1 and 2 in a medium 3 is

$$A_{123} = \frac{3k_BT}{4}\left[\frac{\varepsilon_1 - \varepsilon_3}{\varepsilon_1 + \varepsilon_3}\right]\left[\frac{\varepsilon_2 - \varepsilon_3}{\varepsilon_2 + \varepsilon_3}\right] + \frac{3h\nu_e}{8\sqrt{2}}\frac{\left(n_1^2 - n_3^2\right)\left(n_2^2 - n_3^2\right)}{\sqrt{\left(n_1^2 + n_3^2\right)\left(n_2^2 + n_3^2\right)}\left[\sqrt{\left(n_1^2 + n_3^2\right)} + \sqrt{\left(n_2^2 + n_3^2\right)}\right]} \tag{3.17}$$

The parameters involved are dielectric constants (ε), refractive indices (n), and electronic frequency of absorption (ν_e). The magnitude of the Hamaker constants for materials is in the order of $10^{-19} - 10^{-20}$J and is always positive for particles of the same material. Since Hamaker constants of various materials are available, often a simpler combined or composite approach is used. This entails an estimation based on the geometric mean of the Hamaker constant of the particles and that of the medium as follows:

$$A_{123} = \left(\sqrt{A_{11}} - \sqrt{A_{33}}\right)\left(\sqrt{A_{22}} - \sqrt{A_{33}}\right) \tag{3.18}$$

Based on the DLVO theory, the basis of colloidal stability comes from the sum of the two potential energies from the repulsive electrostatic and the attractive van der Waals forces. Considering spherical particles of the same size, the total potential energy, $V_{\text{tot}}(R)$, is

$$V_{\text{tot}}(R) = \frac{32\pi\varepsilon\varepsilon_0 k_B^2 T^2 \gamma^2}{e^2 z^2} r e^{-\kappa R} - \frac{A r_p}{12R} \tag{3.19}$$

The total potential energy curve (V_{tot}), along with its parts $V_{\text{sph, vdW}}$ and $V_{\text{sph, R}}$ are shown in Figure 3.8(A) for spherical particles. A maximum occurs at a certain separation distance as the repulsive and attractive forces balance each other out. This maximum represents the energetic barrier to colloidal aggregation. Particles must overcome this repulsion barrier to come into contact. At a small separation, attractive interaction dominates, and particles coagulate. Thermal energy (k_BT) can provide sufficient energy to overcome this barrier. At room temperature, $k_BT = 4.1 \times 10^{-21}$J. If the potential energy maximum is below or comparable to this value, the colloidal suspension is not stable. The higher the surface charge density or surface potential, the greater the repulsion and thereby the height of the potential maximum. Keeping all parameters constant, the potential barrier increases with particle size. Another factor that influences this threshold is electrolyte concentration.

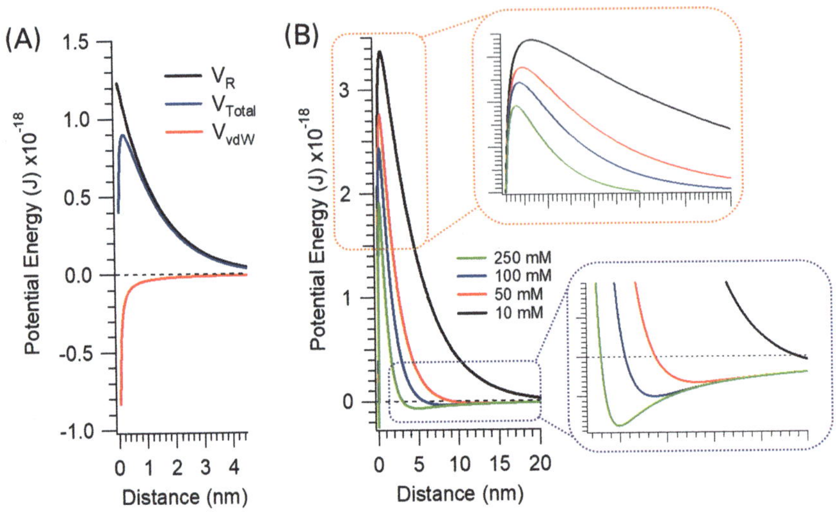

Figure 3.8: (A) The repulsive, attractive, and total particle-interaction potential energy as a function of separation distance. For these curves, $r_p = 50$ nm, surface potential is 100 mV, $A = 1 \times 10^{-20}$J, and the electrolyte concentration is set at 0.1 M. (B) Effect of electrolyte concentration on the total potential energy.

Figure 3.8(B) shows that increasing the electrolyte concentration of the suspension decreases the potential energy maximum. Particles are more prone to aggregation at higher ionic strength. The presence of ions leads to shielding of surface charge, reducing repulsion. Interestingly, based on the DLVO theory, we also observe a secondary minimum at a separation distance larger than the distance at which the maximum occurs. As can be seen, the depth of this minimum is relatively small but larger than room temperature k_BT. Thus, particles can get trapped in this potential minimum. This predicts flocculation of colloids, loosely bound clusters of particles. Unlike coagulation, corresponding to the primary minimum, flocculation is a reversible process. Based on the DLVO theory, it is possible to predict the minimum concentration of electrolyte at which colloidal suspension becomes unstable and coagulates. An expression derived to calculate the critical coagulation concentration (c.c.c) is [58]

$$\text{c.c.c} = \frac{9.85 \times 10^4 (\varepsilon\varepsilon_0)^3 k_B^5 T^5 \gamma^4}{N_A e^6 z^6 A^2} \tag{3.20}$$

For particles with low surface potential, the c.c.c is proportional to z^{-6}, where z is the valency of the counterion. This is known as the Schulze-Hardy rule. The c.c.c value serves as an important parameter for understanding and controlling colloidal stability.

(iii) Repulsion due to steric interaction – particles coated with polymers or other long-chain molecules can repel each other due to steric interactions. This arises from the inability of the interfacial-bound polymer layers of a certain thickness to overlap when two particles come into contact. As the particles approach each other, the local osmotic pressure increases due to steric hindrances. Particles repel each other as the chains compete for the same volume. Another possibility is that as polymer chains overlap, they lose conformational entropy. This loss of entropy is a penalty for the spontaneous conglomeration of the particles. Thus, both entropic and osmotic effects serve as steric repulsion. These effects raise the overall potential energy barrier. The total potential energy can be described as $V_{\text{tot}}(R) = V_R(R) + V_{\text{vdW}}(R) + V_{\text{steric}}(R)$. The mechanisms of steric repulsion dictate that it is relatively insensitive to variations in solution ionic strength and changes in the particle surface potential. To increase steric hindrance, polymers can be attached to colloids via covalent bonding (tethered or anchored chains) or by physical adsorption through hydrogen bonding, hydrophobic interactions, etc. Accordingly, coating particles with polymers to prevent colloids from aggregating is known as steric stabilization. It is a common and important practice in applied colloidal science.

3.6 Adsorption at Colloidal Interfaces

Due to their size range, colloids exhibit large surface-area-to-volume ratios (Section 2.1). Chemical species from the surrounding medium can interact with colloidal surfaces. In this chapter, we have seen that suspended particles can have different interfacial charges. They can also have hydrophobic moieties and different interfacial functional groups providing mechanisms of binding via chemical bond formation. Particle surfaces can be smooth or contain soft interfacial layers. Colloids can also have pores in them. All these factors dictate how molecules from the dispersant interact with and "stick" to the colloidal interfaces. The process of attachment at an interface is called adsorption. The concept of adsorption needs to be distinguished from the processes described by the terms absorption and sorption. Absorption refers to a process where a substance is transferred from the surrounding medium into the bulk volume of another material. Examples include water absorbed by plants and gas molecules absorbed by activated carbon. Sorption encompasses both absorption and adsorption. Adsorption is purely an interfacial phenomenon. The chemical substance (e.g., a gas molecule or a dissolved chemical species) that is adsorbed at the interfacial layer is called an adsorbate. The surface onto which the adsorption occurs is called the adsorbent. If the surface is solid, it is often referred to as a substrate. In biocatalysis, and for thin films, the word substrate has a different meaning, often implying a condensed phase that physically supports a film or layer. Adsorptive is a term given to a substance that has the capacity to be adsorbed.

An adsorbate can attach to the particle surface via strong chemical bonding or through weak electrostatic and van der Waals type-interactions. Adsorption involving chemical bonds is called chemisorption. Physisorption refers to adsorption comprising weak physical forces. Energetically, referring to adsorption from gas phase, sublimation of physisorbed species requires 20–40 kJ/mol of energy, and about 100–400 kJ/mol is needed for chemisorption. Figure 3.9 depicts the difference between these two types of adsorption mechanisms. Species that are physisorbed can freely diffuse at the interface and desorb easily. Physisorbed species do not modify the surface significantly, as they are often located at a greater distance from the surface compared to that of the chemisorbed species. Multilayer formation is possible in physisorption. Chemisorbed species, on the other hand, are relatively fixed. They often restructure the substrate, and do not desorb readily. Chemisorption is highly site-specific due to bond formation and is limited to monolayers. In other words, chemisorption is irreversible and physisorption is reversible if the particle is not porous. For a porous adsorbent, hysteresis may be observed, in which the adsorption/desorption isotherms do not overlap.

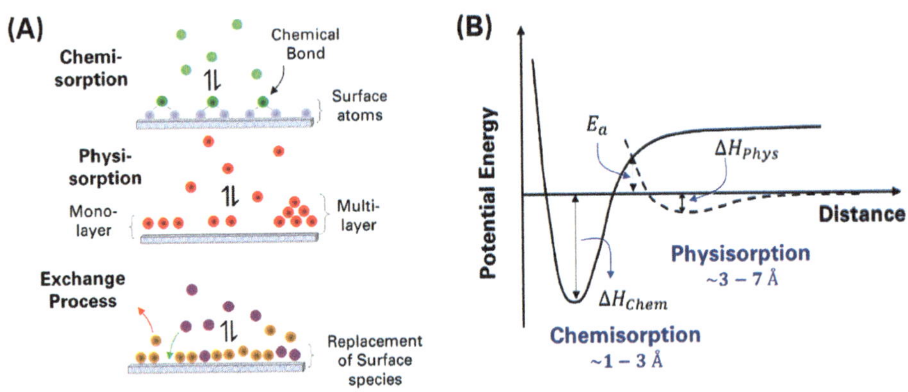

Figure 3.9: (A) Cartoon of different types of adsorption processes. (B) The energetics of physi- versus. chemisorption.

In the case of pure physisorption, there is no activation barrier. Atoms or molecules approaching the surface can readily fall into the physisorption potential energy well. The time to reach adsorption equilibrium is often rapid, whereas the time required to reach equilibrium is slower for chemisorption. This is because formation of chemical bonds involves higher activation energy and complex processes including orientational changes for bonding. The characteristics summarized thus far are applicable for pure physisorption and chemisorption processes. However, for any adsorption process, physisorption is the first step. The weak attractive forces between the substrate and the adsorptive are always present to certain degrees. The species is initially physisorbed, and then it has the potential to overcome an activation barrier to reach the deep energy minimum. The scenario shown in Figure 3.9(B) depicts a substantial barrier to chemisorption, which will influence the kinetics of adsorption. Whether physical or chemical, adsorption from solution can also involve an exchange process.

As shown in Figure 3.9(A), this involves displacement of a different species already bound at the interface with an adsorptive of higher surface affinity. For particles dispersed in water, adsorption of a dissolved organic compound may entail displacing existing water molecules adsorbed at the colloidal/aqueous interface.

3.6.1 Adsorption Isotherm and Kinetic Models

Experimentally, adsorption can be studied by monitoring the amount of a given substance adsorbed as a function of its concentration remaining in the bulk medium. Often, separation techniques are used to assess the change in the adsorptive concentration in the bulk phase before and after the adsorption has occurred. In Chapters 6 and 7, we shall see that laser techniques can allow monitoring of the adsorption process in situ, without separating or perturbing the colloidal suspension by physical force. Irrespective of the approach, the resulting plot of the number (or mass) of adsorbates versus their equilibrium concentration (c) in the solution phase is called an adsorption isotherm. This data is generated at a fixed temperature, hence the term "isotherm". This plot provides the basis for determining the number of adsorbates taken up by the surface, thereby providing information on the particle surface capacity. Thermodynamic parameters such as the adsorption equilibrium constant (K) or similar characteristic parameters can also be obtained depending on the fit model. Table 3.1 provides a collection of common adsorption isotherm models along with their descriptions [59, 60].

Table 3.1: A collection of common adsorption isotherm models.

Adsorption isotherm	Equation	Comments
Langmuir	$\theta = \frac{Kc}{1+Kc}$	Here, K is the adsorption equilibrium constant. At low c or at a low surface coverage, ($\theta \ll 1$), θ varies linearly with c. At $\theta = 0.5$, $K = c^{-1}$.
Multisite Langmuir	$\theta = \sum_i \frac{K_i c}{1+K_i c}$	An extension of the Langmuir model for multiple sites with distinct energetics of adsorption.
Multicomponent Langmuir	$\theta_i = \frac{K_i c_i}{1+\sum_j K_j c_j}$	K_i is the equilibrium constant for the ith species in competition with the jth species. The overall coverage is $\theta = \theta_i + \sum_j \theta_j$.
Henry	$\theta = K_H c$	K_H is Henry's adsorption constant. This model applies to the condition when the surface coverage is very small.

Table 3.1 (continued)

Adsorption isotherm	Equation	Comments
Freundlich	$\theta = K_d c^{\frac{1}{n}}$	K_d is the distribution coefficient, and n, the Freundlich exponent, is usually greater than unity. If $n = 1$, it returns the Henry or linear isotherm. Freundlich model is an empirical equation. It does not show linear behavior at low pressure.
Brunauer-Emmet -Teller (BET)	$\theta = \frac{K_{BET} p/p_0}{(1 - p/p_0)[1 + p/p_0 (K_{BET} - 1)]}$	The term p/p_0 represents relative pressure. It is applicable for physisorption that leads to formation of multilayers. K_{BET} is related to heats of adsorption and condensation. This equation reduces to Langmuir behavior at low p. The BET model is used to measure the pore size distribution of porous materials.
Frumkin-Fowler-Guggenheim (FFG)	$\frac{\theta}{1-\theta} e^{\frac{zw\theta}{k_B T}} = K_L c$	FFG model is a modified Langmuir isotherm that incorporates lateral interaction. The lateral interaction is described by the parameter w, and z is the coordination number. A positive value of w indicates attractive adsorbate-adsorbate interaction.
Hill-de-Boer (HdB)	$\frac{\theta}{1-\theta} e^{\frac{\theta}{1-\theta} - \frac{2\alpha\theta}{k_B T \beta}} = K_{HdB} c$	HdB isotherm model takes lateral interactions into account and allows the adsorbates to be mobile. It is the 2D analog of the van der Waals equation The parameters α and β are the two-dimensional analogs of van der Waals constants a and b, respectively, and are related to the lateral interaction energy.

In the table above, adsorbate quantity is described as a fraction, θ. It is the fraction of the surface sites occupied and is a ratio of the amount (q_{eq}) of adsorbate bound to the substrate at an equilibrium to the possible maximum amount (q_{max}) that can adsorb at the surface. That is, $\theta = \frac{q_{eq}}{q_{max}}$. The adsorbate amounts can be described in the unit of concentrations or number densities. In the latter case, $\theta = \frac{N_{eq}}{N_{max}}$. While many different types of isotherm models have been developed, the Langmuir model has found an extensive application in the realm of adsorption science. It provides a simple and effective method of describing monolayer adsorption on homogeneous interfaces. This model assumes no interactions between adsorbates and that they are immobile or localized at the surface. The following equation describes the adsorption process as a chemical reaction at an equilibrium:

$$c_{eq} + (1 - \theta)S \rightleftharpoons \theta S \tag{3.21}$$

This chemical expression describes the solute concentration, represented by c_{eq}, in equilibrium with the surface, S. At full monolayer coverage, $\theta = 1$, and thus the term $(1 - \theta)$ is representative of the fraction of unoccupied sites. At equilibrium,

$$\frac{d\theta}{dt} = k_a c_{eq}(1 - \theta)S - k_d \theta S = 0 \tag{3.22}$$

Rearranging eq. (3.22) yields the Langmuir equation as shown in Table 3.1, where c represents the equilibrium concentration of the adsorptive. Assuming the forward and reverse processes in eq. (3.22) are elementary steps, we can define $K = \frac{k_a}{k_d}$ as the Langmuir adsorption equilibrium constant, where k_a and k_d are the rate constants for the adsorption and the desorption steps, respectively.

The adsorption kinetic process can be broken down into several stages. The first step is the transfer of the adsorptive molecules from the surrounding solvent to the interfacial region. This is known as the external diffusion process and the concentration gradient is its driving force. Next, the adsorbate diffuses at the interfacial region to the specific site of the adsorbent. This process can involve diffusion through the pores of the adsorbent. Lastly, adsorption (the interaction between the adsorbate and the active sites) occurs. To describe adsorption kinetics, various mathematical models have been proposed and applied. Common adsorption kinetic models [61, 62] are highlighted in Table 3.2. The Langmuir kinetic model is based on the Langmuir isotherm. As seen in the differential equation, not only does θ but also the adsorptive concentration also change with time. As a result, an analytical solution is not feasible, but a numerical method can allow fitting experimental kinetic data. If it is assumed that the change in the concentration in the solution is negligible; that is, it remains the same as the initial concentration, c_0, the integrated form shown in Table 3.2 is achieved.

The pseudo-first-order (or the Lagergren equation) and pseudo-second-order (also known as the Blanchard's equation) kinetic models are widely used. Both are considered empirical models; however, the connection between the Langmuir kinetics model and the pseudo-first-order (PFO) and pseudo-second-order (PSO) models has been suggested. For instance, when the surface coverage is very low or the adsorption rate is much faster than the desorption rate, the Langmuir equation reduces to the PFO model. Whereas, at high surface coverage or when the surface is nearly saturated, the Langmuir kinetics can be approximated by the PSO model. Accordingly, the PFO rate generally describes the adsorption process primarily controlled by the diffusion step and the PSO describes the adsorption on active sites (AAS). A varying-order or mixed-order (MO) model has thus been proposed which theoretically captures both diffusion and active site adsorption processes. The differential equation for the MO model becomes $\frac{dq_t}{dt} = k_1(q_e - q_t) + k_2(q_e - q_t)^2$, which is solved numerically based on Runge-Kutta method. The MO model is thereby applicable to arbitrary initial adsorbate concentration in the solution.

Table 3.2: A collection of common adsorption kinetic models.

Models	Differential equation	Integrated equation
Langmuir kinetics	$\frac{d\theta}{dt} = k_a c_t(1-\theta) - k_d\theta$	$\theta = \frac{k_a c_0 - e^{-(k_a c_0 + k_d)t}}{k_a c_0 + k_d}$ assuming $c_t = c_0$
Pseudo-first-order (PFO)	$\frac{dq_t}{dt} = k_1(q_e - q_t)$	$q_t = q_e\left(1 - e^{-k_1 t}\right)$
Pseudo-second-order (PSO)	$\frac{dq_t}{dt} = k_2(q_e - q_t)^2$	$q_t = \frac{q_e^2 k_2 t}{1 + q_e k_2 t}$
Pseudo-*n*th-order (PNO)	$\frac{dq_t}{dt} = k_n(q_e - q_t)^n$	$q_t = q_e\left(1 - \frac{1}{1 + (n-1)q_e^{n-1}k_n t^{\frac{1}{n-1}}}\right) n \neq 1$
Ritchie's equation	$\frac{d\theta}{dt} = \alpha(1-\theta)^n$	$\theta = 1 - [1 + (n-1)\alpha t]^{\frac{1}{1-n}}\ n \neq 1$
Elovich	$\frac{dq_t}{dt} = \alpha e^{-\beta q_t}$	$q_t = \frac{1}{\beta} ln(1 + \alpha\beta t)$
Boyd's external diffusion	$\frac{dq_t}{dt} = R(q_e - q_t)$	$q_t = q_e\left(1 - e^{-Rt}\right)$

Empirically, adsorption kinetic profiles can be fit using the pseudo-*n*th-order model, however, specific physical meanings based on the kinetic fitting parameters are lost. This *n*th order term refers to kinetic models where the rate of adsorption is proportional to the power function of the amount of adsorbate remaining to be adsorbed, but the exponent (n) does not necessarily correspond to the stoichiometric reaction order. In contrast, Ritchie's equation uses the dimensionless occupancy fraction of adsorption sites (θ) instead of the adsorbed quantity (q_t) at time t. The physical meaning is that adsorption can occur at n active sites and desorption is negligible. The term α describes the adsorption rate constant. Although Ritchie's equation had originally been developed for a gas-solid surface, it has been applied to other heterogeneous surfaces. The Elovich model also considers heterogeneous surfaces and assumes that the activation energy increases with adsorption time. The parameters α and β are constant. The general application of this empirical model has been for chemisorption of gases on heterogeneous solid surfaces.

To recap, AAS can be modeled using the PSO, Langmuir kinetics, Ritchie's equation, and the phenomenological AAS models. The empirical Elovich and pseudo-*n*th-order models are also applied, but the corresponding parameters lack physical meaning. Various models also exist to describe mass transfer processes. For instance, Boyd's external diffusion equation, Fursawa and Smith, Matthews and Weber, and phenomenological external mass transfer models have been derived to describe adsorption where external diffusion of the adsorbates in the surrounding liquid film around the adsorbent is the slowest step. On the other hand, the internal diffusion models assume the rate at which an adsorbate diffuses into the pores of an adsorbent is the slowest step. Examples of internal diffusion models include the Boyd's interparticle diffusion, Weber and Morris, and the phenomenological internal mass transfer models. Additionally, the pore volume and surface diffusion model is used to describe adsorption into porous material. Both external and intraparticle diffusions are considered in this mathematical model. Explicit differential equations representing a unique chemical adsorption process can also be devised. The corresponding integrated forms of the differential equations or numerical analysis can be done to fit the experimental data.

Whether it is adsorption isotherm or kinetic data, these mathematical models, be it empirical or derived considering physical mechanisms in mind, provide a way to understand the adsorption process. Fitting the experimental data often requires a nonlinear regression analysis or numerical analysis. Thus, assessment of the fitting quality is a big part of the data analysis. At a minimum level, parameters obtained from the fitting can be compared for different adsorbates or adsorbents to gain a relative insight into the adsorption process. At best, meaningful physical information such as adsorption strength, maximum adsorbate capacity, rate of adsorption, and mechanisms of the adsorption processes can be gained. Many of these models were originally derived for adsorption processes occurring on planar gas-solid surfaces, however, found applications for colloidal interfaces.

End-of-Chapter Problems

3.1 Concepts and definitions:

(A) Describe the physical meaning of EDL. Compare the concept of Stern layer, Helmholtz layer, the slipping plane.

(B) What is difference between surface potential and zeta potential?

(C) Define (and provide relevant equations) of the following terms: ionic strength, Debye length, GC model, Grahame's equation, electrophoretic mobility, Donnan potential.

(D) Explain the difference between the Smoluchowski and Hückel limits in the context of Henry's function.

(E) Describe different types of soft colloids.

(F) Describe relaxation effect.

(G) What are the major components of the DLVO theory? Provide descriptions of these forces and their influence on colloidal stability

(H) Define the following terms: physisorption, chemisorption, adsorption isotherm, adsorption kinetics, Langmuir model.

3.2 Consider a colloidal system consisting of silica particles of diameter 1 μm dispersed in an aqueous solution of 0.01 M NaCl at 25 °C. The surface potential is measured to be −75 mV.
   (A) Calculate the bulk number density of Na^+, Cl^-, and total ions per m^3
   (B) Calculate the Debye length
   (C) At low potential, $\sinh(x) \cong x$. Starting with Grahame's equation, derive a relationship between ψ_0 and σ_0 applicable at low potential.
   (D) Assuming flat surface, estimate the surface charge density in C/m^2. Comment on the validity of this result based on the magnitude of κr.
   (E) Calculate the surface charge density if the particle diameter is 50 nm and curvature plays a role.

3.3 Derive the Hückel's equation (3.7), where $f(\kappa r) = \frac{2}{3}$) by considering a charged particle of radius r, carrying a surface charge Q, and moving under an applied electric field E, with a constant velocity U. The magnitude of the electric force acting on the particle is $F_e = QE$ and the drag or viscous force on the particle is $F_v = 6\pi\eta rU$ (see Section 2.4.2). At equilibrium, $F_e = F_v$. Given that $\kappa r \ll 1$, we can assume that the surface potential $\psi_0 = \frac{Q}{4\pi\varepsilon\varepsilon_0 r}$. Furthermore, assume $\psi_0 = \zeta$.

3.4 The following data [63] are available from a measurement of the zeta potential in an aqueous suspension of kaolin particles at 25 °C:

Particle diameter	[NaCl]	Time for particle movement	Movement of the particle	Voltage applied across a 10-cm cell
0.5 µm	0.003 M	4.6 s	360 µm	200 V

   (A) Calculate the electrophoretic mobility of the particles.
   (B) Provide an estimation of the zeta potential of the particles.
   (C) Hint: Calculate κr and use appropriate $f(\kappa r)$. Mention one example of how the zeta potential can be changed.
   (D) Explain briefly the role of zeta potential in the study of the stability of colloidal dispersion.
   (E) The zeta potential is often of less importance in systems with organic solvents. Explain why. How can kaolin suspension be stabilized in a solution with an organic solvent?

3.5 Van der Waals forces in colloids and calculation of Hamaker constant.
   (A) Starting from the Hamaker equation for two different particles in a medium, derive the following equation for the Hamaker constant, A_{121}, of particles type (1) in a medium (2):

$$A_{121} = \frac{3k_BT}{4}\left[\frac{\varepsilon_1 - \varepsilon_2}{\varepsilon_1 + \varepsilon_2}\right]^2 + \frac{3h\nu_e}{16\sqrt{2}}\frac{\left(n_1^2 - n_2^2\right)^2}{\left(n_1^2 + n_2^2\right)^{3/2}}$$

(B) Based on the information provided, calculate the Hamaker constant of each of the following identical particles in water: alumina, polystyrene (PS), and silica at 25 °C.

Material	**Refractive index (n)**	**Dielectric constant (ε)**	**Effective electronic frequency (ν_e)**
Alumina	1.76	9.2	~1.8×10^{15} Hz
Polystyrene	1.59	2.55	~1.5×10^{15} Hz
Silica	1.46	3.9	~1.9×10^{15} Hz
Water	1.33	80.1	–

(C) Estimate the effective Hamaker constant assuming the following literature values for the Hamaker constants of the pure materials in vacuum:

Material	**Alumina**	**Polystyrene**	**Silica**	**Water**
Hamaker constant (J)	1.5×10^{-19}	6×10^{-20}	6.5×10^{-20}	5.5×10^{-20}

Calculate the van der Waals energy between two spherical polystyrene particles, each with a radius of 500 nm, immersed in water, and separated by a surface distance of 0.1 nm. Use a Hamaker constant of $A = 1.1 \times 10^{-20}$ J. Assuming no electrostatic repulsion, will this system remain colloidally stable at 25 °C?

3.6 DLVO interaction energy.

(A) Show that eq. (3.14) reduces to $\frac{Ar_p}{12R}$ for equal identical particles with radius r_p.

(B) Calculate the total potential energy $V_{\text{tot}}(R)$ for the interaction between two identical spherical colloidal particles at a specific separation distance R, using the given parameters. Assume steric interaction is negligible.

Hamaker constant	$A = 1.0 \times 10^{-20}$ J
Particle radius	100 nm
Separation distance	10 nm
Debye length	10 nm
Effective surface potential	−25 mV
Electrolyte	NaCl
Temperature	25 °C

(C) Express the total potential energy in terms of thermal energy and comment on colloidal stability at 25 °C.

3.7 The Schulze-Hardy rule in colloid chemistry states that the effectiveness of an ion in causing coagulation (precipitation) of a sol is directly proportional to its valency (charge), provided the ion has an opposite charge to that of colloidal particles.

(A) A negatively charged colloidal sol has a c.c.c of 0.1 M NaCl (sodium chloride). Estimate the c.c.c of the colloid when using the following electrolytes:
   i. $MgCl_2$
   ii. $AlCl_3$

(B) Define the term "coagulation value" (or flocculation value). How does it relate to coagulating power based on the Schulze-Hardy rule?

(C) Explain a practical application of the Schulze-Hardy rule in daily life or industry.

3.8 Adsorption studies of two different vapor species A and B onto a solid surface yield the following information:

- Gas A has an enthalpy of adsorption ΔH_{ads} of –15 kJ/mol. The adsorption occurs rapidly at low temperatures (77 K, liquid nitrogen temperature) and multiple layers form on the surface.
- Gas B has an enthalpy of adsorption of –250 kJ/mol. This adsorption is slow at room temperature but significantly increases at 800 K. The process stops once a single, complete layer is formed.

(A) Do these vapors undergo physisorption or chemisorption? Explain your reasoning.

(B) Why does the adsorption of vapor B increase at higher temperatures, whereas physisorption generally decreases with increasing temperature?

(C) Refer to Figure 3.9(B). Describe the path a molecule would likely take at a very low temperature and if the temperature is raised. How does the depth of the chemisorption well compare to the physisorption well, and what does this indicate about the nature of the forces involved?

3.9 Consider the adsorption of two molecules A and B on the same sites on a given surface. Show that the total equilibrium coverage of both A and B is given by:

$$\theta = \frac{K_A c_A + K_A c_A}{1 + K_A c_A + K_B c_B}$$

Assume adsorption is nondissociative, both A and B follow Langmuirian adsorption kinetics, and that each site can only by either A or B.

3.10 Show that for dissociative adsorption within the Langmuir model, the isotherm equation is given by

$$\theta = \frac{\sqrt{Kc}}{1 + \sqrt{Kc}}$$

3.11 Consider the adsorption of an organic compound onto polymeric nanoparticle. The adsorption isotherms, described by Langmuir equation, at three different temperatures are shown.

i. Estimate the fraction adsorbed at 20 °C when the solution equilibrium concentration is 10 μM.

ii. The adsorption equilibrium constant at 50 °C is 4.9×10^4. What is the Gibbs free energy of adsorption at 50 °C?

iii. Derive a general equation, in terms of K, that gives the equilibrium concentration at which the adsorbent is exactly half covered.

iv. Is the adsorption exothermic or endothermic? Explain your answer. Assume the change in entropy is negligible.

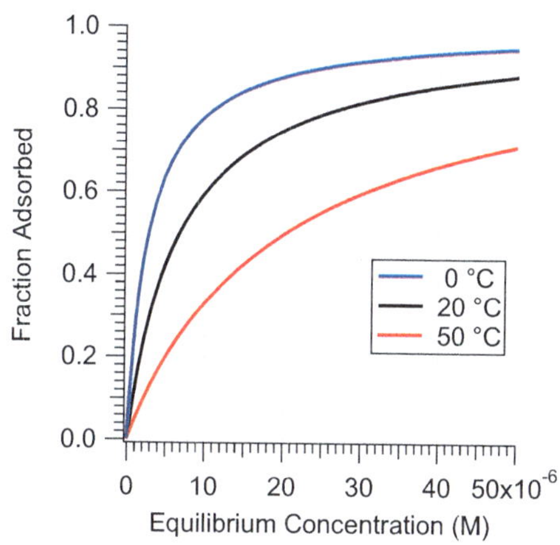

3.12 The following data [63] have been obtained for the adsorption of nitrogen on TiO_2 at 77 K:

p	0.02	0.05	0.1	0.2	0.3	0.4	0.6	0.8	0.9
Vol. of N_2	1.2	1.5	1.9	2.4	2.7	3.0	3.8	4.6	6.7

(A) Make a plot of Vol. of N_2 versus p/p_0 and demonstrate the BET isotherm model represents the data well. Note, $\theta = V/V_{max}$.

(B) Based on the fit, determine the monolayer coverage V_{max} in cm^3/g.

(C) Assuming the area occupied by one nitrogen molecule is $\sigma = 16.2 \times 10^{-20}$ m^2, calculate the specific surface of solid TiO_2 in m^2/g. The specific surface area is given as $A_s = n_{max} N_A \sigma$, where n_{max} is the monolayer capacity in mol/g. The density of N_2 at 77 K is 0.808 kg/m^3.

(D) In a separate study, 0.50 g sample of a spherical mesoporous colloidal silica material was analyzed using N_2 physisorption at 77 K. The specific surface area and the total pore volume were found to be 450 m^2/g and 0.65 cm^3/g. Assuming the pores are uniform, estimate the average pore diameter (d_{pore}). Note, $d_{pore} = 4V_{pore}/A_S$.

4 Light and Colloids: Significance of LASERs

A variety of experimental techniques can be used to determine the fundamental properties of colloids and colloidal interfaces. Generally, measurements require perturbing the system and monitoring the response. Perturbation can be either mechanical force-based or excitation by means of light, electrons, small atoms, ions, and other subatomic particles. The response from the material system can be detected in the form of light emission, scattered subatomic particles (e.g., electrons), or other mechanical responses. It is evident that a wide range of instruments is possible, exploiting different combinations of perturbing the system and detecting the response. In addition to the use of light, a variety of other experimental tools can be used to determine the fundamental properties of colloids and colloidal interfaces. The focus of this book is laser-based techniques. However, it is imperative to be cognizant of the various types of experimental methods that exist for studying colloids. No one technique is perfect. Each has its limitations and advantages. We begin this chapter with an overview of a variety of experimental tools available to characterize colloidal properties. This is followed by a discussion on the fundamental properties of light and, in particular, lasers.

4.1 An Overview of Characterization Methods for Colloids

A summary of the major techniques and the corresponding colloidal properties measured is given in Table 4.1. The analytical techniques listed do not constitute an exhaustive inventory, but these methods have been applied to study various colloidal physical and chemical properties. Moreover, some of these experimental methods capture processes that occur at the surface or the interfacial region of colloids. Many of the experimental techniques in this list involve light in the optical spectral range; that is, visible light as an incident beam and/or as an output that is detected. In light-based techniques, lasers are often a major tool for the source of the incident beam. Other techniques involve physical/mechanical forces (e.g., pressure, gravity, or physical contact) or different forms of radiation (e.g., x-ray) to perturb the system. In many cases, both light-based and other approaches are combined to study colloidal properties.

 | https://doi.org/10.1515/9783111018386-004

Table 4.1: A collection of major analytical tools applied to study colloidal properties [64, 65].

Colloidal property	Common methodologies (techniques)
Size	Light Scattering (DLS, SLS, LD, DCS, EPLS, MALLS, NTA), light obscuration, optical spectroscopy (PLS, UV-Vis), X-ray spectroscopy (XRD, SAXS, EXAFS), microscopy (AFM, SEM, TEM, HRTEM, STM), Coulter principle (TRPS), NMR, FMR, ICP-MS, MALDI, FFF, SEC, HDC, AUC, VIP
Size distribution, polydispersity, or polymodality	DLS, DCS, SAXS, NTA, ICP-MS, FMR, TRPS, SEM, TEM, AFM, STM, FFF, SEC, HDC
Shape or morphology, structural properties	Microscopy (AFM, SEM, TEM, HRTEM, STM, 3D-tomography), light scattering (MALLS EPLS), FMR, SAXS
Molecular weight	DLS, SLS, CD-MS, SEC, GPL with MALLS
Surface charge	DLS, ELS, LDV, PALS (zeta potential), EPM (optical or acoustic), SHG
Colloidal stability, interparticle interaction, and aggregation states	DLS, OLT, FMR, Mössbauer, LBT (surface tensiometer), SEM, TEM, AFM, STM, UV-Vis, DCS, FFF, SEC, HDC, AUC
Elemental analysis and chemical composition	FTIR, NMR, XRD, XPS, SEM-EDS, TEM-EELS, PLS, ICP-MS, STEM, FMR, DSC, MFM, LEIS
Crystal structures	XRD, EXAFS, HRTEM, TEM, electron diffraction
Particle concentration and density	UV-Vis, ICP-MS, CD-MS, PTA, TRPS, DCS, RMM-MEMS
Molecular adsorption	NLS (SHG, ESFG, VSFG), SERS, UV-Vis, BET (measure of porosity), TGA
Interfacial structure and interaction	NLS (SHG, e-SFG, v-SFG, also known as SFS), SERS; QCM-D
Optical properties	UV-Vis, NIR, PLS, EELS-STEM, NLS (SHG, e-SFG)
Magnetic properties	Magnetic properties SQUID, VSM, Mössbauer, MFM, FMR, XMCD, magnetic susceptibility
Rheological properties	Rheometer, viscometer, OLT

List of abbreviations of the major techniques

BET	Brunauer-Emmett-Teller
CDMS	Charge detection mass spectrometry
DCS	Differential centrifugation sedimentation
DSC	Differential scanning calorimetry
DLS	Dynamic light scattering
EDS	Energy dispersive X-ray spectroscopy
ELS	Electrophoretic light scattering
EELS	Electron energy loss spectroscopy

EPM	Electrophoretic mobility
EPLS	Elliptically polarized light scattering
ESFG	Electronic sum frequency generation
FFF	Field flow fractionation
FMR	Ferro-magnetic resonance
FTIR	Fourier transformed IR spectroscopy
GPL	Gel permeation chromatography
HRTEM	High-resolution transmission electron microscopy
HDC	Hydrodynamic chromatography
ICP-MS	Inductively coupled plasma-mass spectrometry
LBT	Langmuir Blodgett Trough
LD	Laser diffraction
LDV	Laser Doppler velocimetry
LEIS	Low-energy ion scattering
MALDI	Matrix-assisted laser desorption/ionization
MALDI-TOF	MALDI-time of flight
MALLS	Multiangle laser light scattering
MFM	Magnetic force microscopy
NIR	Near-infrared spectroscopy
NLS	Nonlinear spectroscopy
NTA	Nanoparticle tracking analysis
OLT	Optical laser tweezers
QCM	Quartz crystal microbalance
QCM-D	QCM with dissipation monitoring
PLS	Photoluminescence spectroscopy
PTA	Particle tracking analysis
RMMM-MS	Resonant mass measurement microelectro-mechanical system
SAXS	Small angle X-ray diffraction
SEC	Size exclusion chromatography
SEM	Scanning electron microscopy
SERS	Surface-enhanced Raman scattering
SFG	Sum frequency generation
SFS	Sum frequency scattering
SHG	Second harmonic generation
SIMS	Secondary ion mass spectrometry
SLS	Static light scattering
SQUID	Superconducting quantum interference device
STM	Scanning tunneling microscopy
STEM	Scanning transmission electron microscopy
TEM	Transmission electron microscopy
TGA	Thermal gravimetric analysis
TRPS	Tunable resistive pulse sensing
UV-Vis	Ultraviolet-visible
VIP	Voltammetry of immobilized particle
VSFG	Vibrational sum frequency generation
XMCD	X-ray magnetic circular dichroism
XPS	X-ray photoelectron spectroscopy
XRD	X-ray diffraction

Broadly speaking, we can categorize these measurement approaches as follows: microscopy (e.g., light, electron, and force), spectrometry, separation, or flow fractionation techniques (e.g., sedimentation, centrifugation, field-flow fractionation, electro- and thermophoresies), spectroscopy (optical, electron, ultrasonic, high-energy, etc.), scattering (diffraction of light, neutrons, and X-rays) techniques, and particle counting or tracking analysis. These different techniques, individually or in combination, can provide information on various colloidal properties, as listed in Table 4.1.

4.1.1 Microscopy

When we hear the term "microscope", most will think about optical microscopy – the type that Zacharias Janssen and Antonie van Leeuwenhoek are credited for inventing. Microscopy, as originally invented, uses light and lenses to magnify small objects. In an optical microscope, light is passed through a condenser to focus it on the sample. As the light passes through the sample, it goes through the objective lens. A magnified image of the sample is then transmitted to the oculars. Thus, a microscope in a sense captures a magnified shadow of the object under study. One can observe this projected image with the naked eye or capture it using an advanced camera. With advanced software, particle movement and tracking can be accomplished. The optical microscope played a pivotal role in visualizing biological cells, which are in micrometer dimensions. In an ideal optical microscope, the lateral resolution (objects side by side at the same image depth perpendicular to the beam) is approximately 200 nm, whereas axial resolution (objects that are separate and distinct when parallel to the beam) is around 500 nm. Thus, optical microscopy has been used to image colloids. In fact, Jean Perrin and his collaborators observed colloidal suspensions of mastic and gamboge particles using a microscope quantitatively to determine Avogadro's number (Section 1.2.1). Quantitative optical and confocal microscopy are now common tools to characterize colloids [66–69]. Microscopes are also coupled with various other techniques including but not limited to infrared (IR), fluorescence, multiphoton processes, light scattering, optical tweezers, and rheology.

Microscopy-based techniques also include SEM, STM, STEM, and TEM. Instead of light, electron beams are used in these techniques. For instance, in SEM, a narrow electron beam of high energy (typically 20–30 keV) is focused onto a solid substrate. The electron has a de Broglie wavelength that is much shorter than typical light waves, and thus yields a much better spatial resolution. This generates highly localized backscattered and secondary electrons, which are detected and processed to construct an image. The number of electrons detected depends on the variations in surface morphology. Raster scanning along the *xy* direction allows 3D mapping of the surface and provides topographic and morphological information of solid particle surfaces with a typical lateral resolution of 1–50 nm. Conventional SEM requires vacuum compatible samples that are occasionally coated with a conducting film. However, within the last decades, the development of environmental SEM and atmospheric SEM has allowed observation of various materials under low or near-atmospheric pressure and without conductive coating. A disadvantage of environmental SEM and atmospheric SEM is a slight loss of resolution.

While in SEM, reflected or back-scattered electrons are detected, TEM uses electrons that are transmitted through the sample. Using TEM, information on the internal structure of the material, such as morphology, stress state, and crystal structure, can be obtained. High-resolution TEM provides much higher magnification (on the order of $15,000–300,000\times$) compared to conventional TEM. In STEM, both dark- and bright-field imaging like TEM are obtained, and at the same time, a focused convergent beam is used to probe the sample surface as in SEM. In addition to the use of light and electrons, other forms of microscopy exist to probe colloidal samples. AFM is a force-based microscopy, whereas STM utilizes quantum tunneling of electrons and tunneling current to image surfaces. These are well-renowned and powerful surface probe microscopic techniques with applications in colloidal science [70–73]. In AFM, a very sharp tip is positioned at the end of a flexible cantilever, which detects changes in force or displacement as the probe tip moves over the sample surface. A laser beam is deflected off the cantilever, and the tip's movement in the *z*-direction, is detected. The displacement of the cantilever occurs due to intermolecular repulsive forces when the tip is brought close to the surface. Imaging both conductive and insulating surfaces can be done using AFM. This technique can be used in various environmental conditions, such as air and liquid, and does not require the sample to be in a vacuum. A lateral resolution of 10 Å or less is achievable using AFM. On the other hand, STM provides a lateral resolution of ~1 Å at the atomic scale and a vertical resolution of ~0.1 Å. However, it is applicable to solid conductors, semiconductors, and insulators with conductive coatings. Both AFM and STM can produce 3D images.

In summary, microscopy techniques can yield local information about the particles along with their size, shape, dispersity, aggregation state, growth kinetics and mechanism, particle morphology, and surface characterization. The application of optical light microscopy is rapid and simple. It is also cost-effective and routinely found in laboratories. However, it does not provide good spatial resolution for nanomaterials or colloids with dimensions in the ~sub-500 nm range. Electron microscopy provides far superior resolution and greater magnification. Some disadvantages are that this equipment can be expensive and require samples to be in conditions such as vacuum or ultralow pressure. A common disadvantage for these methods is that colloidal samples often need to be deposited on a substrate, and the particles are not in a suspended state. This precludes in situ investigation of airborne particles or soft matter dispersed in liquid suspensions. Both SEM and AFM have the advantage of mapping out surface morphology of colloids. Additional and detailed reviews of the application of microscopic techniques can be found elsewhere [66–73].

4.1.2 Mass Spectrometry

There is a subtle difference between the terms “spectrometry” and “spectroscopy.” According to the IUPAC definition, the latter involves the study of physical systems by the electromagnetic radiation (EMR) with which they interact or that they produce. In contrast, spectrometry is the measurement of radiation as a means of obtaining information about the systems and their components. Although these terms are often used interchangeably, spectrometry methods do not entail a direct measurement of the interaction of the chemical system with light. Whereas spectroscopic techniques derive molecular information from the radiation itself upon interaction with the system. We shall discuss spectroscopic and light-scattering techniques in subsequent sections. Here, we outline a major spectrometric method, namely mass spectrometry, that has been applied to study colloidal substances. Mass spectrometry in general uses an ionization source to generate ions, and the mass-to-charge (m/z) ratio is determined. It is considered a spectrometric technique because the fragmentation pattern is used to analyze the molecule rather than a direct measurement of the interaction of the molecule with EMR. There are several variations of mass spectrometers. ICP-MS, single particle ICP-MS, MALDI-TOF, SIMS, and CDMS have been applied to characterize colloidal nanomaterials [74–76].

ICP-MS is an elemental analysis technique. In general, the sample is ionized using inductively coupled plasma (typically, argon plasma) in which energy is produced by electric current. Thereafter, the mass analyzer separates and detects ions based on their m/z ratio. ICP-MS can provide information on elemental composition and size distribution of particles. In SIMS, secondary ions are analyzed using a mass spectrometer based on their m/z ratio. It provides high spatial resolution and surface sensitivity when analyzing colloids. In MALDI-TOF, a laser is used to desorb and ionize molecules. Then, the time it takes for the ions to travel, which correlates to the m/z ratio of the ions, is measured. While commonly used for characterizing metallic nanomaterials, advances in these techniques allow mass spectrometry to be applied to a wide range of colloids.

4.1.3 Fractionation Techniques

Fractionation refers to the process of separating a mixture into different components. Thus, separation or fractionation methods combined with other means of selectively detecting different characteristics (e.g., size, charge, and chemical composition) of colloids provide a valuable approach in determining colloidal properties. There are several ways to separate colloidal mixtures such as centrifugation, filtration, sedimentation, hydrodynamic chromatography, and field flow fractionation. In field flow fractionation, colloids are passed through a thin channel under an applied field [77, 78]. Examples of applied fields include gravitation, electric fields (e.g., gel and capillary electrophoresis), thermal fields, centrifugal fields, hydraulic fields, or crossflow. The latter refers to applied fields perpendicular to the main flow through a channel. Separation is achieved when the interacting particles susceptible to that force accumulate at the channel walls. Noninteracting particles move forward. Multiple detectors can be used to track the continuous separation and the fractionated compositions. Scattered light (including UV-Vis and fluorescence) or ions are commonly detected.

4.1.4 Spectroscopic and Light-Scattering Techniques

The method of investigating physical and chemical systems using EMR is the basis of spectroscopic techniques. EMR can be described as a wave, or a particle (photon). In the wave theory, electromagnetic (EM) fields are characterized by their intensity and frequency or wavelength. In particle theory, photons carry energy, and their energy is proportional to their frequency. The higher the frequency, the greater the energy of the EMR. The range of frequencies encompassing all types of EMR is known as the EM spectrum. Matter, on the other hand, is composed of electrons, protons, and neutrons, and they occupy different discrete energy states. Therefore, when EMR of different frequencies interact with material systems, it can evoke different responses. For instance, ultraviolet (UV) or visible light provides sufficient energy to cause electrons to jump from a lower energy state to a higher one. Excitation is associated with absorption spectroscopy. If the transition is from a higher to a lower energy state, and energy is released in the form of light, the method is called emission spectroscopy.

The frequency of light absorbed or emitted is unique to the atomic or molecular transition. As a result, both absorption and emission spectroscopic techniques are powerful techniques to identify elements, molecules, or chemical functional groups within a colloidal substance. For example, XPS provides information on elemental analysis, chemical state, and bonding for colloidal surfaces. IR-based techniques such as FTIR and attenuated total reflection are applicable to probe chemical composition based on vibrational modes of deposited colloidal substances. Techniques based on nonlinear spectroscopy (NLS), such as SFG and SHG, are used to gain interfacial information of various colloids in situ. The intensity of the light absorbed or emitted is often correlated with the amount of material present in the sample. Thus, spectroscopy provides quantitative analysis of chemicals in colloidal matter as well.

In addition to absorption and emission, light can scatter off matter. Scattering is a process in which light is deflected from an obstacle; in other words, light changes direction as it interacts with particles. It occurs primarily when the wavelength of the light is comparable to the size of the particles. Since colloidal substances fall in the size realm of 1 nm to 1 μm, UV, visible, and NIR EMR are effective in scattering light from colloids. UV light can induce photodecomposition and NIR radiation can cause thermal degradation in materials. Due to its strong overlap with colloidal size range, visible light (380–780 nm, ca.) is a common source for scattering techniques such as

SLS and DLS. They provide information on colloidal size and molecular weight. SERS is another technique that uses enhanced Raman scattering.

A specific case of scattering is diffraction. When light interacts with a material, the direction of the light wave changes due to interference effects. The outcome is the spreading of the wave as it passes through an opening or a slit. The diffracted waves create a distinct pattern of maxima and minima, where light intensities are added constructively and destructively, respectively. Diffraction is observable when the slit width is comparable to or smaller than the wavelength of the light. In addition to the wavelength and incident angle of the light, the diffraction pattern also depends on the size and shape of the diffracting object. Thus, light diffraction is a useful tool to obtain information about colloidal properties such as particle size, size distribution, and shape. Details of light scattering and spectroscopy, in conjunction with application of lasers, are discussed in Section 4.4.

4.1.5 Additional Techniques

Colloidal properties can be obtained based on force measurements. For example, rheological properties of colloids such as viscosity and elasticity are measured using rheometers. Rheology is the study of deformation and flow of matter. Measurement of viscosity (material's resistance to flow) and elasticity (ability to return to its original shape) helps understand and control the behavior of colloidal systems in various applications. A rheometer is an instrument used to measure the way materials respond to applied force. It often relies on force measurements. Analyzing the force and deformation allows determination of viscosity, elasticity, shear, and stress.

Surface tension is another property measured based on force. We have seen (Section 2.5.1) that there are a variety of approaches to measure interfacial tension. LBT is a technique that allows controlled compression and expansion of monolayers of deposited samples generally at the air/aqueous interface. In this case, adsorbed species are insoluble and form a Langmuir monolayer upon compression. Unlike a Gibbs monolayer, when a Langmuir monolayer is compressed, the molecules do not desorb into the subphase. Surface tension is measured based on the Wilhelmy plate method (Section 2.5.1) while the monolayer spread is compressed. The difference in the surface tension of the pure liquid subphase, γ_o, and that of the subphase with the monolayer, γ, is called the surface pressure, π. As the barriers are closed slowly, the area (A) that the molecules occupy decreases. This causes an increase in surface pressure. A plot of π versus A isotherm is essentially a two-dimensional phase diagram of the monolayer. An analogy is a pressure versus volume plot for three-dimensional movement of gases at a fixed temperature. The parameter A in this case is the mean molecular area, and it is calculated based on the trough area and the moles of the insoluble monolayer deposited. Although LBT has been commonly used to study lipid and surfactant molecules, nanomaterials and colloidal substances have also been investigated recently. The π versus A curve provides information on particle-particle interactions, self-assembly, aggregation state, and phase transitions.

Additionally, investigation of molecular adsorption onto the particle surface also yields various information on colloidal properties. For example, BET isotherms provide information about the porosity of the particles. As discussed in Section 3.6.1, various mathematical models are available to fit adsorption isotherm data. These isotherm profiles not only help elucidate thermodynamics of molecular binding but also provide information about particle surface area. Adsorption isotherms for suspended particle-molecule systems can be generated by means of separation techniques or in situ using NLS. In either case, light is often used to probe adsorption processes.

4.2 Fundamentals of Light-Matter Interaction

Many of the techniques surveyed in Section 4.1 utilize light within the optical range, which includes UV, visible, and IR radiation. Light is a common ingredient and can be used either directly or indirectly. Table 4.1 highlights some of these methods. One way light can be used is to invoke a response in the colloidal system. Another is to detect the light emitted, transmitted, or scattered from the material under investigation. The third is the combination of both or in conjunction with a separate probing mechanism. It is unmistakable that light-matter interaction, a subject we review in Section 4.3, is vital in understanding colloids and processes involving colloidal interfaces. But a brief primer on the basic properties of light is necessary.

4.2.1 Wave Description of Light

Light can be described as EM transverse waves or as particles. In the wave picture, the EM field oscillates and travels in space and time. The electric ($\vec{E}$) and the magnetic ($\vec{B}$) fields are orthogonal to each other as shown in Figure 4.1(A). All EM waves obey Maxwell's equations and propagate at a speed of 2.998×10^8 m/s, also known as the speed of light, c. In vacuum, $c = (\varepsilon_0 \mu_0)^{-\frac{1}{2}}$, where ε_0 and μ_0 are the electric permittivity and magnetic permeability in free space, respectively. In a medium other than a vacuum, the speed of the EM wave is $\frac{c}{n}$, where n is the refractive index of the medium. The wave characteristic entails a disturbance in space and time. The wavelength (λ) of the EM wave is the distance between two successive crests or troughs. The frequency (ν) of the oscillation is related to the wavelength and the speed of light as follows: $c = \nu\lambda$. Additional terminologies pertinent to describing EM waves include the wavevector $\vec{k}$ and its angular frequency (ω). The magnitude of the wavevector is given as $\left|\vec{k}\right| = \frac{2\pi n}{\lambda}$ and $\omega = 2\pi\nu$, where n is the refractive index of the medium.

Maxwell equations – a set of four fundamental equations that describe how changing electric and magnetic fields propagate through space – describe how the shape of an EM wave evolves over time and space. They capture the way disturbances propagate through a medium. Concise vector forms of the wave equations are shown in eqs. (4.1a) and (4.1b). They are second-order partial differential equations that relate the second derivative of the field displacement with respect to time (right-hand side) with the second derivative of the displacement with respect to position (left-hand side). The solutions to these equations are the wave functions that describe the specific behavior of the wave. The mathematical operator ∇^2 determines the curvature of a function in space and is given as $\nabla^2 = \frac{\partial^2}{\partial x^2} + \frac{\partial^2}{\partial y^2} + \frac{\partial^2}{\partial z^2}$

$$\nabla^2 \vec{E} = \varepsilon_0 \mu_0 \frac{\partial^2 \vec{E}}{\partial t^2} \tag{4.1a}$$

$$\nabla^2 \vec{B} = \varepsilon_0 \mu_0 \frac{\partial^2 \vec{B}}{\partial t^2} \tag{4.1b}$$

Figure 4.1(A) depicts a light wave propagating in the $x-$ direction, designated by the propagation wave vector $\vec{k}$, and its $\vec{E}$ field oscillating in the $z-$ direction. This figure represents a linearly polarized or plane-polarized light. The orientation of the $\vec{E}$ field is constant in one direction even though its sign and magnitude vary over time. Due to the transverse nature of the wave, there is no $\vec{E}$ field component in the direction of its propagation. Mathematically, a harmonic or sinusoidal function, which satisfies the wave equation, is useful in describing the waveform. In general, the overall electric field is the sum of its vectorial components as follows: $\vec{E} = E_x + E_y + E_z$. Based on Figure 4.1(A), we note that $E_x = 0$ (direction of the propagation) and that there is no component in the y-direction; that is $E_y = 0$. Thus, this waveform can be described by eq. (4.2a):

$$E_z(z,t) = E_{0z} \cos[\omega t - kx] \tag{4.2a}$$

$$E_y(y,t) = E_{0y} \cos[\omega t - kx + \phi] \tag{4.2b}$$

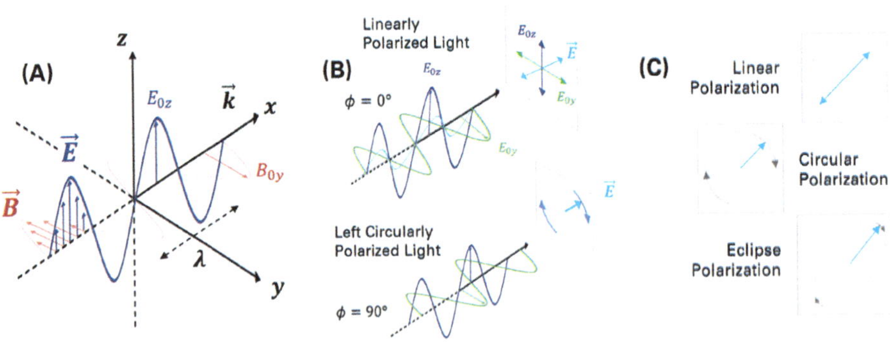

Figure 4.1: (A) Electromagnetic wave. (B) Depiction of polarized EM fields and (C) their representation in polar plot.

Given that the field is vibrating only along the *z*-direction (direction of gravity), we say this wave is linearly polarized along the *z*-direction. It is vertically polarized or parallel to the direction of gravity. Linear polarization refers to a situation when the electric field oscillates in a single plane along the direction of polarization. Now, consider there are two different electric fields, both propagating in the *x*-direction, as shown in Figure 4.1(B), top panel. One oscillating along the *z*-direction (on *xz*-plane) as before, and the other oscillating along the *y*-axis (*xy*-plane). The latter is horizontally polarized, meaning perpendicular to direction of the gravity. In relation to the former, this horizontally polarized light has the form expressed in eq. (4.2b). The term ϕ is introduced to describe the relative phase difference between the waves. It indicates how much one wave is shifted relative to another. For the specific waveform shown in this example, $\phi = 0$. These waves are aligned peak-to-peak and are considered in phase. Additionally, their amplitudes are equal, i.e., $E_{0z} = E_{0y}$. If these two waves are added, the resultant wave has its electric field $\vec{E}$ oscillating along the $z = y$ plane. This new wave, as shown (Figure 4.1(B), top), is also linearly polarized but tilted at 45° from the *z*-axis.

The example provided above is a special case. In general, two waves need not have the same amplitude and can be out of phase. Circular polarization occurs when the electric field of the light consists of two linear components that are perpendicular to each other, equal in amplitude, but have a phase difference of 90° (Figure 4.1(B), bottom panel). The resulting electric field rotates in a circle around the direction of propagation. Left circularly polarized light has an electric field vector that rotates counterclockwise when viewed along the direction of propagation, while right circularly polarized light has an electric field vector that rotates clockwise. Elliptical polarization results from the combination of two linear components with different amplitudes and/or a phase difference that is not 90°. This is the most general description of polarized light, whereas circular and linear polarized light are special cases of elliptically polarized light. Figure 4.1(C) summarizes these polarization configurations.

In contrast to polarized light, unpolarized or nonpolarized light refers to light where the electric field oscillation direction is random and constantly changing, with no single dominant orientation. Natural light (e.g., sunlight) is unpolarized. This means the light waves oscillate in random directions because the light source is comprised of numerous individual emitters. The radiated EM fields are independent from each other. Lasers, on the other hand, produce strongly polarized light. Natural light can however become partially polarized due to reflection or scattering. Partially polarized light refers to light waves where the electric field oscillations are not completely aligned in a single direction but instead have a preferred direction with a stronger intensity compared to others. Materials can also rotate polarized light in different directions. The amount of rotation is determined by the chiral nature, and the concentration of the molecules. This is the premise for identifying which stereoisomer may be present in a sample.

Lastly, it must be noted, EM waves carry energy. The average energy per unit area per unit time is known as the irradiance (I) or intensity. The irradiance is proportional to the square of the amplitude of the electric field as follows:

$$I = \frac{\varepsilon_r \varepsilon_0 c}{n} \left\langle E_0^2 \right\rangle_T \tag{4.3}$$

The bracket, $\langle \ \rangle_T$, represents the average value over time.

4.2.2 Particle Description of Light

Not all light-matter interaction phenomena can be explained using the wave properties of light. Experimental observations, such as the photoelectric effect and spectral lines, require describing light as a particle. This light particle, known as photon, is essentially a discrete energy packet, the energy of which is related to the frequency of the light as follows: $E_{\text{photon}} = h\nu$, where h (6.626×10^{-34} Js) is the Planck's constant. The overall energy of N number of photons is $Nh\nu$. Photons are massless and travel at a speed of light, c. Within the particle picture of light, the polarization of light is a result of the intrinsic spin of the individual photons. A photon is an example of a bosonic particle and thus has a spin of 1. Circularly polarized light consists of photons with only one type of spin. Linearly polarized light, on the other hand, is composed of photons that are in a superposition of both right and left circularly polarized states. In conclusion, light exhibits both wave-like and particle-like behavior, known as the wave-particle duality. Sometimes it manifests as a wave, while other times it acts as a particle. We shall toggle between the two in describing light-matter interactions as appropriate.

4.3 Light-Matter Interaction: Optical Spectroscopy

When light interacts with matter, light energy can be absorbed by the atoms or molecules that make up the matter. Additionally, the light can get scattered or diffracted off the material. In this section, we focus on the former and the scattering process in Section 4.4. Light absorption induces motion at the atomic or subatomic scale. These motions can include movement (vibration) or breakage of bonds, transition of electrons from a lower energy state to a higher one, or ionization, increased rotational motion, or change in the nuclear spin state. The type of molecular event that occurs is dependent on the energy of the photon, which in turn is related to the frequency of the light. Thus, different information is obtained with the different EM waves used. They are summarized in Table 4.2. Optical spectroscopy refers to UV, visible, and infrared (IR) radiation. Vibrational and rotational excitations in a molecule are caused by IR and microwaves, respectively. UV-Vis radiation causes electronic transitions in atoms or molecules.

Table 4.2: Spectroscopy based information.

Spectral region	Frequency (s^{-1})	Motion and energy level transitions	Information obtained
γ-ray	$10^{19} - 10^{21}$	Nuclear transitions	Mössbauer spectroscopy – electronic and magnetic environments of specific isotopes, oxidation states.
X-ray	$10^{17} - 10^{19}$	Core electrons $\sigma \leftrightarrow \sigma^*$ transitions	X-ray absorption spectroscopy (XAS), X-ray fluorescence (XRF), X-ray photoelectron spectroscopy (XPS) – electronic structure of materials, elemental composition, and chemical states.
Visible, UV	$10^{14} - 10^{17}$	Valence electrons $\pi \leftrightarrow \pi^*, n \leftrightarrow \pi^*$ transitions	UV-Vis absorbance, fluorescence, phosphorescence spectroscopy – electron configuration, molecular identification, concentration determination.
IR	$10^{12} - 10^{14}$	Vibrational (nuclear motion)	IR spectroscopy – functional group identification, bond strength.
Microwave, far-IR	$10^{9} - 10^{12}$	Rotational	Bond lengths, bond angles, and dipole moments.
Radio waves	$10^{7} - 10^{9}$	Nuclear spin states	Nuclear magnetic resonance (NMR) spectroscopy – functional group identity, electronic environment of the nucleus.

Figure 4.2(A) is a molecular energy diagram that highlights the electronic, vibrational, and rotational energy levels, and selected transitions. A fundamental concept is that these energies are discrete or quantized. Rotational energy states are embedded within a vibrational state whereas multiple vibrational energy levels occupy a given electronic state. The key takeaway is that the difference (ΔE) in the energies of the final and initial energy levels (also known as the transition energy) is equal to the energy of the light, i.e., $E_{\text{final}} - E_{\text{inital}} = E_{\text{photon}}$. Expressing the photon energy in terms of the wavelength of the light, we get $\Delta E = \frac{hc}{\lambda}$. Thus, a spectrum is a plot of the amount of light absorbed (or emitted) as a function of wavelength (or frequency) of the light.

When an electron absorbs light and occupies an excited state, eventually it falls back to a lower energy state. In the process, it can release energy in the form of light. If the transition is from an excited singlet state to a ground singlet state ($S_1 \rightarrow S_0$), the emitted light is called fluorescence. A singlet state refers to a state where all electron spins are paired. If the transition is from a higher triplet state to a singlet state ($T_1 \rightarrow S_0$), the emitted radiation is known as phosphorescence. A triplet state refers to a state where two electrons in different orbitals have parallel spins. Not only absorption of light but also emission of light provides valuable information about the electronic structure of molecules. Both absorption (microwave, IR, UV-Vis) and emission (fluorescence, phosphorescence) processes form the basis of molecular spectroscopy. Fluorescence and phosphorescence are examples of spontaneous emission. A

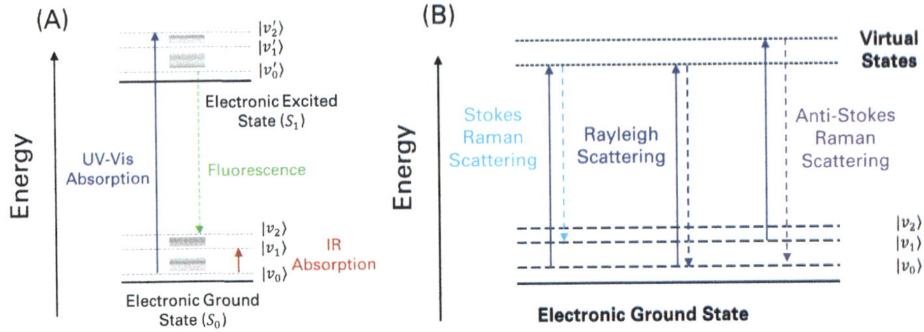

Figure 4.2: (A) Molecular energy diagram highlighting discrete electronic (e.g., S_0 and S_1), vibrational (e.g. $|v_i\rangle$), and rotational energy levels. Electronic transitions corresponding to UV-Vis absorption and fluorescence are shown. Vibrational transitions are caused by IR light absorption. (B) Different types of light scattering represented in the form of energy diagram.

special type of emission known as stimulated emission, the mechanism by which the laser works, is discussed in Section 4.5.1.

4.3.1 Absorption and Turbidity

A spectrum is a graph in which the *y*-axis generally shows how much light has been absorbed (or emitted), and the *x*-axis is the wavelength (frequency or photon energy) of the light. Up until now, the discussion was about the *x*-axis – the origin of the wavelength at which an absorption or emission peak is observed. It is associated with the energy of transition in molecular and atomic systems. Now we turn to the *y*-axis. The value of absorbance signifies the attenuation of light by the matter. Consider light with intensity I_0 passing through a sample of pathlength l. The light that emerges on the other side has an intensity (I). Generally, $I < I_0$ because some of the light has been lost due to absorption or scattering (Tyndall effect). The ratio $\frac{I}{I_0}$ is called the transmittance (T) and absorbance is $-\log_{10} T$. Furthermore, for a sufficiently dilute solution, it can be shown that

$$A = \varepsilon c l \tag{4.4}$$

This is called the Beer-Lambert law, where c represents the concentration (mol/L) of the absorbing species and ε (not to be confused with dielectric constant) is the molar absorptivity or molar extinction coefficient. The unit of molar extinction coefficient is $M^{-1}cm^{-1}$. The extinction coefficient is related to the molecular cross-section due to absorption, σ, which describes the effective area that a molecule presents to the photon flux:

$$\sigma = \frac{2.3\varepsilon}{N_A} \tag{4.5}$$

For electronic transitions, the intensities (or the absorbance value) can be calculated using the concept of oscillator strength, (*f*), which is a dimensionless quantity that describes the probability of an electronic transition in a molecule or atom. It is directly related to the molar extinction coefficient and the integrated sum of the absorption or emission band as follows:

$$f = 1.44 \times 10^{-19} \int \varepsilon(\nu)d\nu \tag{4.6}$$

The integration describes the quantity of area under a curve of the molecular extinction coefficient plotted against frequency. Assuming a Gaussian spectral profile, we have $\int\varepsilon(\nu)d\nu \approx \varepsilon_{\max}\Delta\nu$. The classical concept of oscillator strength is related to the quantum mechanical concept of transition dipole moment, $\vec{\mu}_{12}$, as follows:

$$f_{12} = \frac{8\pi^2 m_e \nu}{3he^2} |\vec{\mu}_{12}|^2 \tag{4.7}$$

For an allowed transition from state 1 to 2, f_{12}~1, and for a forbidden transition, $f_{12} \ll 1$.

The Beer-Lambert law is applicable when the solution is homogeneous, the incident light is monochromatic, and there are no significant chemical interactions between the solute and the solvent or among solute molecules that affect absorption. That is, eq. (4.4) applies to dilute solutions. For most UV-Vis spectrophotometers, it is practical to measure an *A* value between 0.001 and 3. When light passes through a solution containing no particles larger than 1 nm, the path of the light is said to be optically clear. However, in the presence of colloids, light is scattered, which is known as the Tyndall effect, named after John Tyndall's discovery in 1871. The path of light that is visible through the medium due to this scattering is called a Tyndall beam. A sunbeam caused by dust particles is a primary example of a Tyndall beam.

In addition to absorption, scattering also weakens the intensity of the light passing through a sample. Like eq. (4.4), the absorbance in this case is described as $A = \tau l$, where τ represents turbidity, which is a measure of the cloudiness or haziness of a fluid caused by suspended particles. If we combine attenuation of light due to electronic transition and turbidity, we find

$$I = I_0 e^{-(\varepsilon c + \tau)l} \tag{4.8}$$

The reciprocal of τ is the length of the turbid medium required to decrease the intensity of a light beam to $\frac{1}{e}$ of its initial intensity. For spherical macromolecules, it was shown by Lord Rayleigh in 1871 that turbidity is related to the molar mass of the macromolecule. Thus, it is possible (see Chapter 5) to determine the molecular weight of colloidal substances based on light-scattering techniques.

In the absence of vibronic coupling, we can further understand the linear optical properties of typical organic molecules by considering the following relationship:

$$\varepsilon = n^2 = 1 + 4\pi\chi^{(1)} \tag{4.9}$$

The $\chi^{(1)}$ is called the first-order susceptibility and it is related to the dielectric constant and the complex refractive index of the material. We shall see later (Chapter 6) that the induced polarization in a medium is related to the susceptibility terms. The first-order susceptibility is responsible for linear spectroscopy and has the following general expression:

$$\chi^{(1)} = \frac{N}{\hbar}\sum_e \left[\frac{\mu^i_{eg}\mu^j_{ge}}{\omega_{eg} - \omega - i\Gamma_{eg}} + \frac{\mu^j_{eg}\mu^i_{ge}}{\omega_{eg} + \omega + i\Gamma_{eg}}\right] \tag{4.10}$$

In this equation, Γ represents the damping term that accounts for the dissipation of energy during the polarization response, ω_{eg} is the transition frequency, and ω the frequency of the electric field. In addition, μ^i_{eg} represents the ith Cartesian component of the transition dipole moment vector from ground state (g) to excited state (e). For the case where $\Gamma = 0$, there is no absorption, n is real and corresponds to the refractive index of the medium. The situation where $\Gamma \neq 0$ corresponds to optical absorption.

4.3.2 Significance of IR and UV-Vis Spectroscopy for Colloidal Study

Transition energies are distinct for different materials and therefore a spectral signature can serve to identify elements, molecules, and functional groups present in a compound. Measuring the amount of light absorbed or emitted can allow the determination of concentrations of chemical species. The spectral positioning is often sensitive to the local environment; thus, a shift in wavelength in the presence and absence of chemical or EM perturbation also yields information on molecular interaction. With respect to colloids, UV-Vis absorption and fluorescence spectroscopy have been applied to study quantum dots and colloidal nanoparticles (NPs). For instance, metallic gold and silver particles exhibit a phenomenon known as surface plasmon resonance, which results in a characteristic absorption peak in the UV-Vis spectral range. Detection of this peak confirms the formation of these NPs. These particles also exhibit wavelength shifts with a change in particle size. Typically, the absorbance maximum shifts to a longer wavelength as particle size increases. Moreover, spherical NPs generally exhibit a single surface plasmon resonance peak, while rod-like particles yield two peaks, corresponding to the transverse and longitudinal plasmon resonances. Thus, UV-Vis spectroscopy can be useful in estimating the size and shape of certain types of nanocolloids.

Absorbance measurements also provide a way to monitor colloidal aggregation and stability. When particles aggregate over time, a shift in absorbance value or broadening of the peak can occur. The Beer-Lambert law can also be applied to quantify the concentration of certain NPs exhibiting a distinct absorption peak. UV-Vis spectroscopy has been employed to investigate the interaction between colloids and other molecules with absorption bands in the UV-Vis region. For instance, the adsorption of proteins onto the surface of NPs can induce changes in the absorption spectrum, which can be monitored by this method. A separation method can also be employed to study adsorption using UV-Vis of a target compound.

While UV-Vis light causes electronic transitions, IR light excites vibrational modes, which describe the ways molecules can vibrate. A linear molecule can vibrate $3N-5$ ways, whereas a nonlinear molecule has $3N-6$ vibrational modes. The N represents the number of atoms in the molecule. Each vibration mode has a specific frequency (how fast it oscillates) and thereby is excited by a specific frequency of IR light. Based on their unique vibrational signatures, IR spectroscopy can readily identify different functional groups that are present in colloids. It is a common practice to describe the vibrational transition energy in the unit of wavenumber (cm^{-1}). Vibrational transitions are observable if there is a change in the dipole moment of the molecule during vibration. Thus, if a chemical bond has a permanent dipole moment, it will be IR active. If it does not, it can still be IR active if the vibration causes a change in the dipole moment. In addition to this gross selection rule, specific selection rules dictate that $v = \pm 1$ transitions are the most intense, and overtone bands corresponding to $v = \pm 2, \pm 3, \ldots$ transitions are weak. Chemical bonds in organic compounds exhibit bending and stretching modes in the range of approximately 400–4,000 cm^{-1}.

Figure 4.3 provides an IR signature map for common chemical bonds. Ligands attached to the particle surface, and thus the surface functionalization of NPs can be studied using IR absorption spectroscopy. For interfacial selective information, vibration sum-frequency generation is a powerful tool (Chapters 6 and 7). Similar to UV-Vis, monitoring vibrational signatures, IR-based techniques also serve to study colloidal stability and molecular interactions.

4.4 Light-Matter Interaction: Light Scattering

There are many forms of light scattering. Generally, scattering can be characterized as elastic or inelastic scattering. When the wavelength of the scattered light and the incident light are the same, it is called elastic scattering. In this case, the energy of the photons is not transferred to the object, or vice versa, but their direction is changed. Chief examples of elastic scattering include Rayleigh scattering and Mie scattering. In contrast, Raman scattering is inelastic. Energy is transferred when light interacts with matter and the incident photon either loses or gains energy upon collision. Most of the scattered light we see daily can be understood based on the Rayliegh and Mie scattering theories. We consider their mathematical description below. In essence, Rayleigh theory is applicable to understanding light scattering from particles much

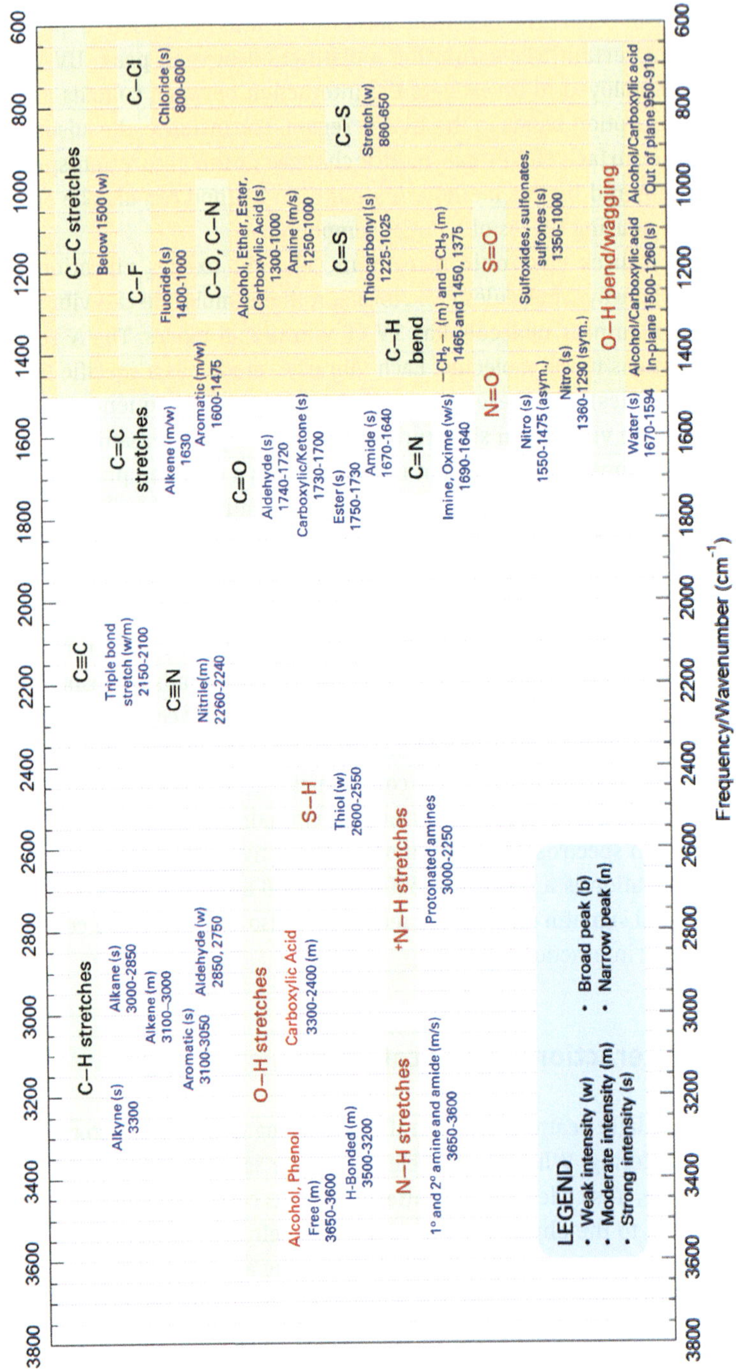

Figure 4.3: IR frequency chart correlating absorbance wavenumber for stretches and bending of specific chemical bonds. The shaded area between ~1500 and ~400 cm^{-1} is known as the figerprint region.

smaller than the wavelength of the light. This applies to scattering from small molecules such as O_2 or N_2 and explains why the sky is blue. Rayleigh scattering strongly favors shorter wavelengths. Mie scattering occurs when the particles are similar in size or larger compared to the wavelength of the light. A key element of Mie scattering efficiency is that it shows a lower dependence on the wavelength. That is, particles that fall under the Mie scattering regime scatter all visible wavelengths nearly equally. As such, clouds or fog, which are much bigger than the wavelength of the visible light, appear white.

The intensity of the scattered light depends strongly on the size, shape, and composition of the particle. It also depends on the wavelength of the radiation and the direction in which the light is scattered. A convenient way to understand the relationship between particle size to the wavelength of the scattered light is to define a size parameter, $x = \frac{2\pi r}{\lambda}$. It is effectively a ratio of the particle radius to the wavelength of the light. Rayleigh scattering theory is applicable when $x \ll 1$ and Mie scattering theory is necessary when $x \sim 1$. Based on this parameter, a plot of particle size (diameter) versus EMR wavelength, in the range of UV to microwave, is shown in Figure 4.4. Colloidal size range is highlighted by the shaded rectangle, which shows that colloids scatter UV, visible, and near-IR light. Size parameter lines for $x = 2,000,\ 0.2,\ \text{and}\ 0.005$ are shown. Mie scattering theory is applicable for a wide range of size parameters. Based on the size parameter thresholds, it can be seen (Figure 4.4) that Rayleigh theory is not suitable to describe scattering from colloids but from atoms and small molecules. However, an approximate solution to light-scattering theory known as the Rayleigh-Gans-Debye (RGD) approximation can be applied to understand scattering from small colloids, assuming that the relative refractive index of the particle is close to that of the surrounding medium. When x is very small, scattering is negligible, and when it is very large, geometric optics, which describes light propagation in terms of rays, can be used.

4.4.1 Rayleigh Scattering

Figure 4.4 reveals that Rayleigh scattering captures only a small portion of the colloidal domain. Very small particles can display Rayleigh scattering. Nevertheless, it is of pedagogical interest to review this theory. Let us consider an unpolarized light wave of wavelength λ impinged upon a small spherical particle of radius r and refractive index n. If the particle radius is much smaller than the wavelength, the intensity of the light scattered (I_s) from one particle in relation to the intensity of the incident beam (I_0) is then given as

$$\frac{I_s}{I_0} = \frac{8\pi^4 r^6}{\lambda^4 R^2}\left(\frac{n^2-1}{n^2+2}\right)^2\left(1+\cos^2\theta\right) \tag{4.11}$$

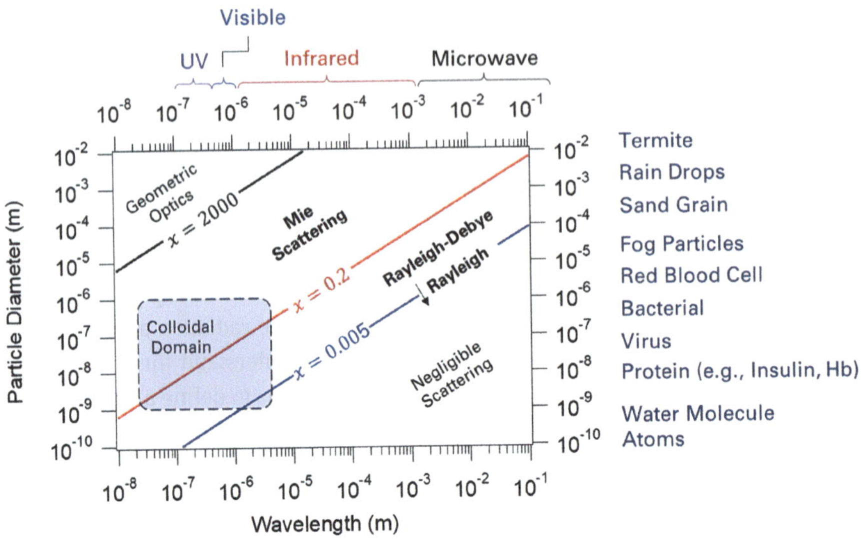

Figure 4.4: Relationship between particle size and wavelength of light, and the applicability of scattering theory.

The parameters R and θ describe the observer's distance from the source and the scattering angle, respectively (see Figure 4.5(A)). The scattering intensity drops according to the inverse-square law. There are three key points: (1) within the Rayleigh regime, the intensity increases by r^6; larger particles scatter light very strongly. (2) Scattering intensity is highly sensitive (raised to the fourth power) to the wavelength of light. For instance, blue light (say, 380nm) will be scattered more effectively than red light (e.g., 750 nm). Figure 4.5(B) shows how the intensity varies with wavelength when detected at 0° and 90° angles. This is the basis of why the sky during the daylight looks blue. The sunlight (white light) contains all wavelengths of the visible spectra, but the scattering of the high-frequency blue color from molecular oxygen and nitrogen is most intense. (3) The scattering intensity depends on the angle at which it is detected. Figure 4.5(C) shows a three-dimensional profile of the scattering intensity and a polar graph showing its projection on a two-dimensional equatorial plane. In this figure, unpolarized light is incident along $\theta = 0^\circ$ and hits the scatterer at the origin. For particles with size parameters much smaller than 1 ($x \ll 1$), the scattered light shows wavelength dependency, and scatter all around the source, including backscattering. There is anisotropy in the intensity with respect to the direction it is measured. In the forward direction, $\theta = 0^\circ$, the scattered light is also unpolarized and its intensity is maximum. At $\theta = 90^\circ$, the light is vertically polarized and has half the intensity relative to the forward direction.

Rayleigh's equation can also be expressed in terms of molecular polarizability and molar mass of the scatterer (eqs. (4.12) and (4.13)). Molecular electronic polariz-

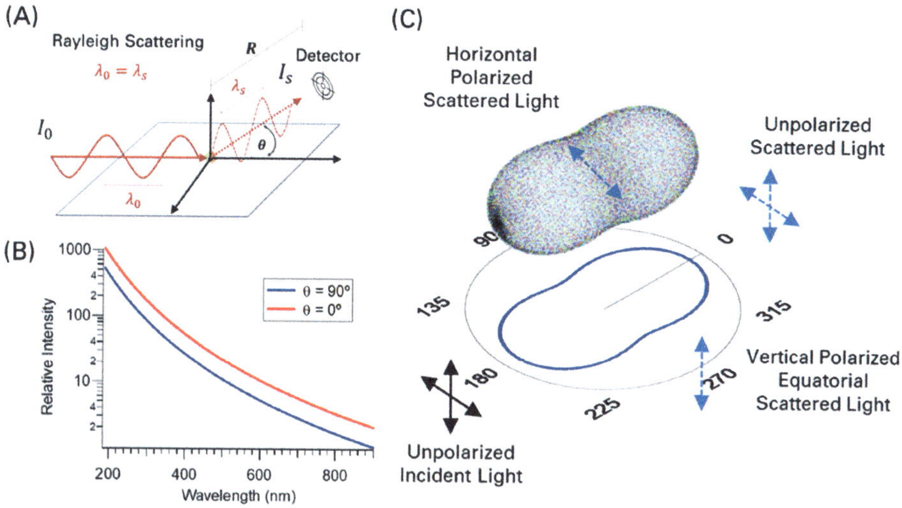

Figure 4.5: (A) Schematic of the scattering geometry. (B) Relative intensity of the scattered light as a function of wavelength at two different detection angles based on Rayleigh model. (C) Three-dimensional scattering intensity profile.

ability (α) describes the ease at which the electron cloud of a molecule is distorted by an external electric field, such as an EMR, leading to the formation of an induced dipole moment. Because electrons are much lighter than the protons and neutrons, the EM field pulls electrons more than it repels the nucleus and thereby causes the molecule to be polarized. In the case of light scattering with linearly polarized light, the electrons are polarized along the direction of the oscillating EM field. This occurs on the timescale of femtoseconds (10^{-15} s). We shall see later that α is related to molecular electronic susceptibility (χ), which in turn will be useful in discussing NLS (Chapter 6). At this juncture, it suffices to say that α is related to molar mass. This is because larger molecules with more electrons are easily distorted by an electric field, leading to a higher degree of polarization. Such a dependence on the molar mass (M) is shown in eq. (4.12). This serves as the basis for measuring molecular weight of macromolecules using laser light-scattering methods (see Chapter 5):

$$\frac{I_s}{I_0} = \frac{\pi^2 N_A \rho \alpha^2}{2\varepsilon_0^2 \lambda^4 R^2 M}\left(1+\cos^2\theta\right) \tag{4.12}$$

$$\frac{I_s}{I_0} = \frac{2\pi^2 M}{\lambda^4 R^2 N_A \rho}(n-1)^2\left(1+\cos^2\theta\right) \tag{4.13}$$

An extension to the Rayleigh theory is the RGD approximation, applicable for weakly scattering or "soft" particles. It is valid when the refractive index of the particle (n_p) is like that of the surrounding medium (n_s). In other words, $|m-1| \ll 1$, where m is the ratio of particle refractive index to that of the surrounding (($m = n_p/n_s$). This situation results in a small reflection of the incident wave at the particle surface and suggests that particles scatter light only weakly. Another assumption is that $2x|m-1| \ll 1$, where is the size parameter. This condition prevents a large change in the amplitude and phase of the incident wave inside the particle. As with Rayleigh theory, the RGD solution also displays wavelength- and angular- dependence of scattered intensity as follows:

$$\frac{I_s}{I_0} = \frac{c\Delta\rho^2 V_p^2 V_s P(q) S(q,c) f(\theta)}{R^2} \tag{4.14}$$

The parameters are: c, number concentration of the suspension; V_s, scattering volume; V_p, particle volume; R, detection distance from the source; and the scattering length density $\Delta\rho = \frac{3\pi n_s^2}{\lambda^2}\frac{m^2-1}{m^2+1}$. The geometrical factor $f(\theta)$ accounts for the polarization of the incident and detected light. The functions $P(q)$ and (q,c) are factors accounting for intra- and interparticle interference, respectively. The scattering wave vector, q, plays a critical role in all light-scattering processes, and is further discussed in Chapters 5 and 6. As an example, the form factor of a spherical particle is

$$P(q,r) = \left[\frac{3}{(qr)^3}(\sin(qr) - qr\cos(qr))\right]^2 \tag{4.15}$$

The functional form of this equation reveals an interference pattern as qR becomes greater than 1. The RGD model has been applied to study small colloids of arbitrary shape and of slightly larger in size than that of the Rayleigh size limit. As highlighted in Chapter 6, this theory has also been applied to nonlinear scattering of light from molecules adsorbed at colloidal surfaces.

4.4.2 Mie Scattering

In 1908, Gustav Mie developed a complete physical description of the scattering of EM plane waves based on the exact solution to the Maxwell equation for homogeneous spherical particles. Unlike the Rayleigh model, Mie theory applies to all possible size parameters. It is valid for all sizes of spheres, indices of refraction, and wavelengths. The scattered light is represented as an infinite series of fields of spherical multipoles that are orthogonal at a sphere surface. However, Mie theory is mathematically complex, and thus, Rayleigh's equation is used for the $x \ll 1$ size range and RGD theory for small colloids. In this section, a qualitative description of the Mie scattering is provided. Scattering profiles based on Mie scattering theory are displayed in Figure 4.6.

Mie theory calculations involve determining the efficiency of scattering, which relates the cross-sectional area of scattering, σ_s, to the cross-sectional area of the spheri-

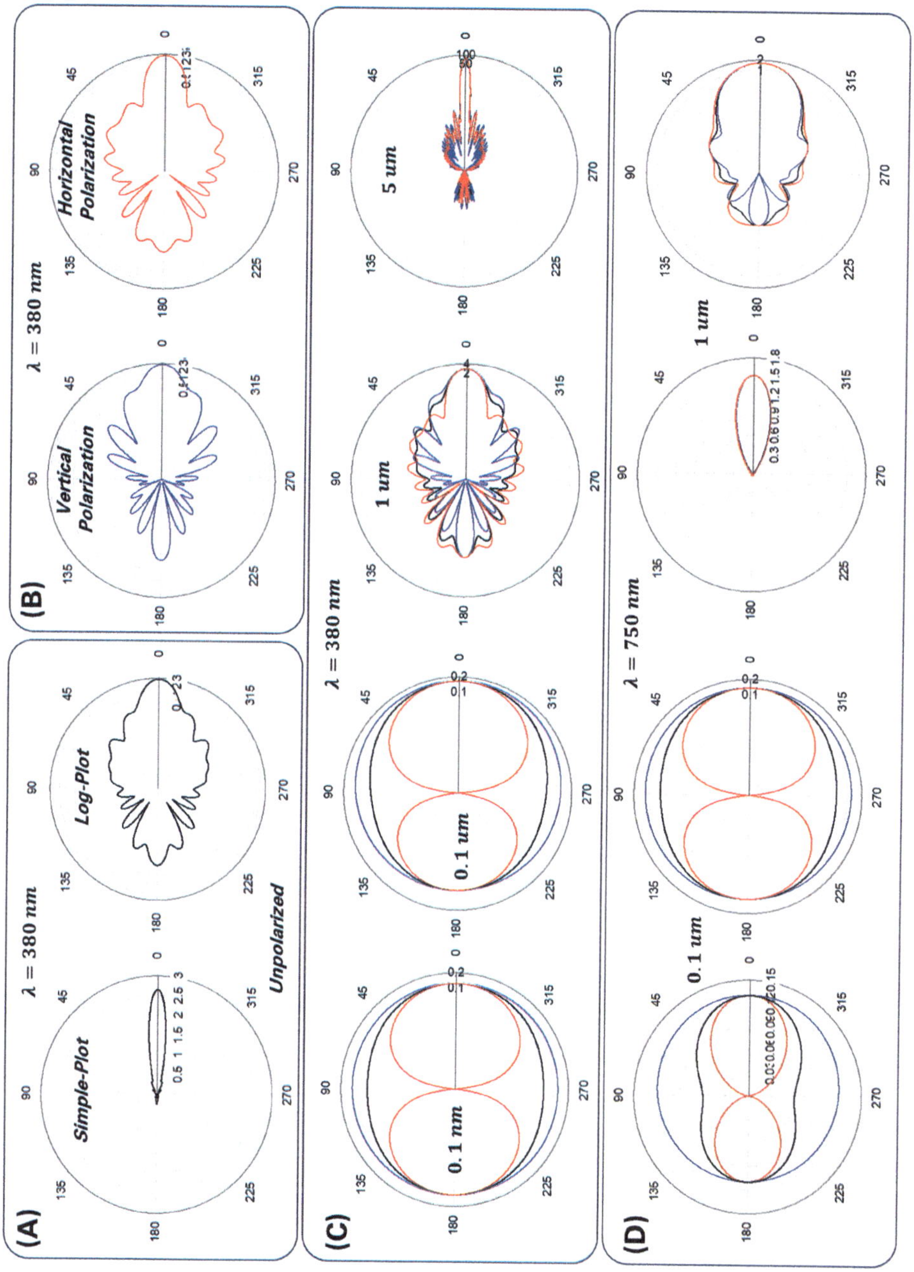

Figure 4.6: Polar plots of Mie scattering from spherical particles under different conditions. Light is incident from the left. (A) Unpolarized Mie scattering from a 1 um particle shown as simple versus log-plot. In the log-plot, the radial distance is plotted on a log scale, with radius origin set at 0.001. (B) Log-plots with the same conditions in (A) but vertical versus horizontal polarized scattering profile shown separately. (C) Effect of particle size on Mie scattering shown as log-plots. For (A)–(C), the wavelength of

cal particle (A) as follows: $\sigma_s = Q_s A$. At the heart of the Mie theory is the scattering efficiency, Q_s, which is expressed (eq. (4.16)) in terms of complex coefficients a_n and b_n, representing electric and magnetic multipoles:

$$Q_s = \frac{2}{x^2} \sum_{n=1}^{\infty} (2n+1) \left(|a_n^2|^2 + |b_n^2|^2 \right) \tag{4.16}$$

First, we consider Mie scattering of blue light ($\lambda = 380$ nm) from spherical particles of 1 μm diameter (Figure 4.1(A,B)). In this scenario, x~8.3. In the simple, linear plot, it shows that scattering of light occurs predominantly in the forward direction. But when plotted in log-scale, backscattering, as well as small side lobes, comes to view. It is clear that Mie scattering produces intricate patterns with multiple lobes, maxima, and minima in the angular distribution of the scattered light. Moreover, the pattern is also different for the different polarizations of the light detected (Figure 4.6(B)). Certainly, Mie scattering profile is different compared to Rayleigh scattering. This is because light is scattered from larger particles. Due to light scattering from different parts of the particle with varying phases, multiple scattering events and interference effects take place. In the Rayleigh case, the small size of the scatterer (relative to the wavelength of light) preserves the isotropic nature of scattering.

The effect of size on the scattering pattern is observed in Figure 4.6(C). When the particle is very small (0.1 nm diameter) compared to the wavelength of the light, we see a pattern resembling Rayleigh scattering emerge. For particles much larger than the wavelength (e.g., 5μm), scattering in forward direction dominates, resembling rays of light emanating from the scatterer. This example demonstrates that geometric optics (raytracing) become relevant in describing light propagation through larger objects. In the intermediate size range, intricate patterns are observed, and the use of Mie theory is a must. In conclusion, Mie theory is exact and requires extensive computation. Rayleigh theory is not applicable for most colloids, and RGD has some usefulness, but has its limits.

4.4.3 Raman Scattering

The type of scattering we have been discussing so far involves elastic collision of photons with materials. The energy of the photon is not altered. In an energy diagram (Figure 4.2(B)), Rayleigh scattering is represented as an interaction where a photon (shown as the up arrow) is absorbed by a molecule, and then immediately re-emitted

Figure 4.6 (continued)
the incident light is 380 nm. (D) Scattering profile when the wavelength is 750 nm for particles with 0.1 and 1.0 um diameters. Both linear and log-plots are shown. For these simulations, refractive indices of the medium and the particle are set to 1.0 and 1.5, respectively. Particle density is $1{\times}10^8$ mL^{-1}. Intensity calculations were performed using the OMLC Web-Based Calculator [79].

(down arrow) with the same energy. The molecule is momentarily in a "virtual state" before returning to its original energy level. A virtual state is a short-lived ($<10^{-15}$s for an energy difference of 1 eV), unobservable quantum state. It is an imaginary intermediate state of a system that provides mechanistic understanding of multistep quantum processes. Now, let us consider scattering due to the inelastic collision. In Raman scattering, the scattered photon has a different energy due to the molecule transitioning to a different vibrational energy level. The frequency of incident light is shifted, and the shift corresponds to the vibrational transition energy. The scattered photon can either lose (Stokes shift) or gain (anti-Stokes shift) energy as shown in Figure 4.2(B). The latter is less probable, resulting in weaker intensity than that of the Stokes shift. Stokes lines are generally more intense because of relatively higher population of molecules in the ground vibrational state. In a Raman spectrum, a Stokes shift appears at a longer wavelength (lower frequency) than the incident light, while anti-Stokes at a higher frequency. The Raman shift (in wavenumber) is calculated as follows:

$$\Delta\tilde{\nu}\left(\mathrm{cm}^{-1}\right) = \frac{1}{\lambda_{\mathrm{laser}}(\mathrm{nm})} - \frac{1}{\lambda_{\mathrm{scatter}}(\mathrm{nm})} \times 10^{7}\left(\frac{\mathrm{nm}}{\mathrm{cm}}\right)$$

In this equation, λ_{laser} is the wavelength of the incident beam (often a laser) and $\lambda_{\mathrm{scatter}}$ is that of the scattered light. The selection rule for vibrational Raman spectroscopy is different from that of the IR absorption spectroscopy. For a Raman peak to be observable, there must be a change in the molecular polarizability during vibration. As described earlier, polarizability represents the ease of distorting electron clouds from their original position. The polarizability of a molecule increases with decreasing electron density, decreasing bond strength, and increasing bond length. As a result, some vibrations are active in Raman but not in IR, and vice versa. This makes IR and Raman techniques highly complementary for obtaining a complete picture of a molecule's vibrational spectrum. Accordingly, they are both valuable in analyzing colloidal chemical compositions.

4.5 LASER and Its Significance

LASER is an acronym which stands for Light Amplification by Stimulated Emission of Radiation. The application of lasers is ubiquitous such that the word "laser" is often used as a common noun in everyday language. It is available in everyday consumer products such as CD/DVD/Blu-Ray players, barcode scanners, laser printers, laser pointers, certain smartphones, and high-speed internet systems. In these examples, the laser power is relatively weak (<5 mW) and therefore nonhazardous. Indeed, high-power lasers are used in scientific research settings, as well as in industries. For example, lasers are used in medical applications such as eye surgery, hair/tattoo removals, and even in dental procedures. In manufacturing, high-precision cutting and welding metals and plastics, and microfabrication require lasers. High-speed data transmission using fiber optics and communication with satellite also utilize lasers.

4.5.1 Fundamentals of LASER

While there are a wide variety of lasers, the basic mechanisms of how lasers work are the same. Three key processes unique to lasers are population inversion, stimulated emission, and amplification. In Section 4.3, we have seen examples of spontaneous emission such as fluorescence and phosphorescence. As shown in the top panel of Figure 4.7(A), the emitted photon can travel in any direction and have an arbitrary phase. This happens when an atom or molecule transitions from a higher energy state (e.g., excited state labeled $|2\rangle$) to a lower energy state (e.g., ground state labeled $|1\rangle$), releasing energy in the form of light. Spontaneous emission is a random process and occurs even in the absence of an external EM field. That is, this type of radiation is incoherent. In contrast, stimulated emission is the process in which an excited atom radiates light under the influence of an incident photon. The result is two photons with identical properties. They are in phase, have the same polarization, and travel in the same direction (Figure 4.7(A), bottom panel). Stimulated emission is coherent and is a critical mechanism for the operation of lasers.

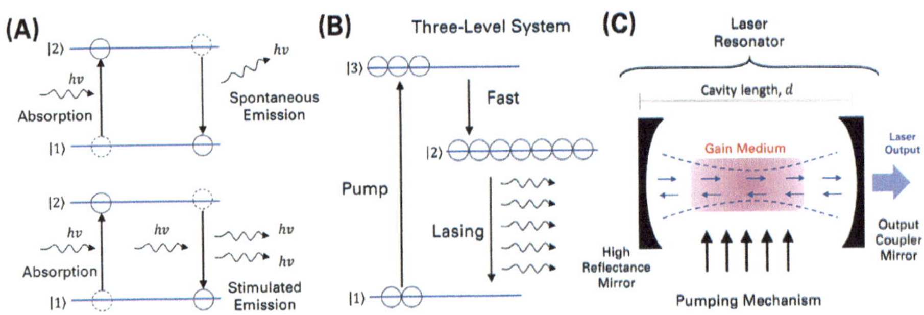

Figure 4.7: (A) Comparison of spontaneous and stimulated emission. (B) Three-level system showing population inversion and lasing. (C) Basic features of a laser resonator.

The contrast between spontaneous and stimulated emissions portrayed in Figure 4.7 (A) is depicted with a two-level system involving only the ground and the excited states. Unfortunately, stimulated emission in a two-level system does not lead to "lasing"! The verb lasing refers to the generation of a coherent laser beam. This is because in a two-level system in thermal equilibrium, the lower energy level will always have a higher population than the upper level. Absorption will dominate, preventing the buildup of atoms or molecules occupying the higher energy level. The situation where the upper energy state has a greater population than that of a lower energy state is called population inversion, and this is a must for lasing to occur. A two-level system cannot sustain a continuous population inversion. At least three energy levels are needed to allow a population inversion. A three-level system is shown in Figure 4.7 (B), highlighting population inversion in energy level $|2\rangle$. A process known as pumping allows excitation to the energy level $|3\rangle$ and if the transition from $|3\rangle$ to is fast, a buildup of level $|2\rangle$ population occurs as level $|1\rangle$ is depleted. A four-level system is more efficient because the lower energy level is already depleted, requiring less energy to pump atoms to the higher level and sustain the inversion. Therefore, less energy input is required to maintain lasing, resulting in a more efficient and continuous operation. In most lasers, materials with typically three or four energy levels are utilized to create the necessary conditions for stimulated emission and lasing. The first successful laser developed by Maiman in 1960, the ruby (Al_2O_3:Cr^{3+}) laser, is a three-level system. It emits a deep red light of wavelength 694.3 nm. Examples of four-level lasers include Nd-YAG, He-Ne, CO_2, and Ti:Sapphire (Ti^{3+}:Al_2O_3) lasers.

Another fundamental aspect of laser operation is the amplification of emitted light. For this, a resonator is required. A schematic of an optical laser resonator or cavity is shown in Figure 4.7(C). It includes a gain medium, bounded by highly reflective mirrors that keep the light from escaping. The photons in the laser cavity thus pass through the gain medium, bounce back and forth, and gain energy. One of the mirrors is called the output coupler, which is a partially transparent dielectric mirror and serves to leak part of the built-up photons (optical power) to produce the output laser beam. The gain medium, which can be solids (e.g., crystals or glasses doped with ions), liquids (solution of organic dyes), gases (e.g., Ar, He-Ne, and CO_2), or semiconductors, is the material where stimulated emission occurs. In general, the gain medium can be pumped optically or electrically. Electrical pumping is common for gas and semiconductor-based lasers. In gas lasers, this is achieved by passing an electrical discharge or high voltage across electrodes through the gas medium, causing the gas to ionize and generate free electrons. Free electrons accelerate, collide with other atoms in the gas, and excite them to achieve population inversion. In semiconductor lasers, applied current causes electrons to move across the semiconductor junction to create electron-hole pairs. Upon recombination of the electrons and holes, photons are emitted, which are amplified by stimulated emission. Optical pumping uses light to excite atoms in the gain medium to achieve population inversion. For example, the Ti:Sapphire laser can be pumped by an argon-ion or frequency doubled Nd:YVO_4 green laser generating light in the range of 514–532 nm.

There are many different laser cavity designs. The shape of the laser beam is determined by the type of resonator cavity. In scientific research, the most common laser beam shape is Gaussian, in which the intensity distribution follows a Gaussian distribution. That is, the intensity drops with increasing distance from the center of the beam. It provides the most compact beam, with optimal focusing properties and minimum possible divergence. The cavity geometry and inclusion of different optical components within the resonator influence the characteristics and performance of lasers significantly. The resonator geometry can impact the laser output power, the cavity stability (minimizing losses and generating consistent output), beam divergence (the spread of the beam), wavelength selection, and mode structure. There are two types of modes: longitudinal and transverse. The former refers to the standing wave patterns along the laser beam path within the cavity (laser's optical axis) and is related to the cavity length. It dictates the wavelength and monochromaticity of the laser output. The transverse mode in contrast describes the spatial distribution or the shape of the laser beam and its ability to maintain focus. Essentially, it defines the laser beam's intensity in the plane perpendicular to the direction of propagation. The fundamental or the lowest order transverse mode of a laser beam is TEM00 or a Gaussian-shaped beam.

The output laser beam can be either a continuous wave or pulsed. Optoelectronics such as mode-locking and Q-switching devices are used in pulsed lasers. Q-switching uses an electro-optic or acousto-optic modulator to rapidly switch the optical losses within a cavity, enabling generation of short and high energy laser pulses. Mode-locking can be either active or passive. The former uses a modulator to control the laser's gain or loss; whereas, in passive mode-locking, a saturable absorber (a material whose absorption decreases with increasing light intensity) is utilized. The passive mode-locking mechanism can generate shorter pulses than active mode locking because the saturable absorber can respond faster to intensity changes than an electronic modulator. The key idea is that pulsed beam in the order of nano- to femtoseconds can be achieved. The term "ultrafast laser" is used for lasers producing short pulses less than a picosecond. Attosecond pulses have also been generated using phase locked high-order harmonic generation. There also exists quasi-continuous wave lasers, which produces pulsed laser beams but with durations of micro- to millisecond range.

4.5.2 Properties of LASER Light Source

Indeed, the mechanism of light generation is different for a laser than other light sources. Lasers offer many advantages such as coherence, high intensity, directionality, focusability, inherent polarization, monochromaticity, and pulsed beam. This is due to the stimulated emission process, which ensures that the emitted photons have the same polarization and phase as the incident photons. It is the latter that is important in determining the coherent nature of the laser light source. Figure 4.8(A) compares the phase difference in incoherent and coherent waves. The laser cavity confines and amplifies this radiation while maintaining the phase over time and distance. The coherence length (L_c) of a light source can be estimated from its central wavelength and its spectral linewidth or bandwidth ($\Delta\lambda$) as follows: $L_c \approx \frac{\lambda^2}{\Delta\lambda}$. It describes the distance over which a light wave maintains a predictable phase relationship. Laser light sources have a much greater coherence length than ordinary lights.

Coherence, coupled with amplification, results in high intensity in lasers compared to the ordinary light source. Stimulated emission and the design of the resonator dictates the directionality of the laser beam. Only the light waves propagating along the cavity's axis are bounced back and forth multiple times and undergo amplification. Waves traveling at an angle are lost through the sides of the cavity and are not amplified significantly. Light from an ordinary light source spreads whereas the amplified photons from a laser travel in the same direction, Figure 4.8(A). Due to their coherence and directionality, lasers are highly focusable. The term "focusability" refers to the ability to concentrate laser light into a small, intense spot. The quality of the focused beam ultimately depends on various factors including the wavelength and the beam diameter of the laser, and the optics used.

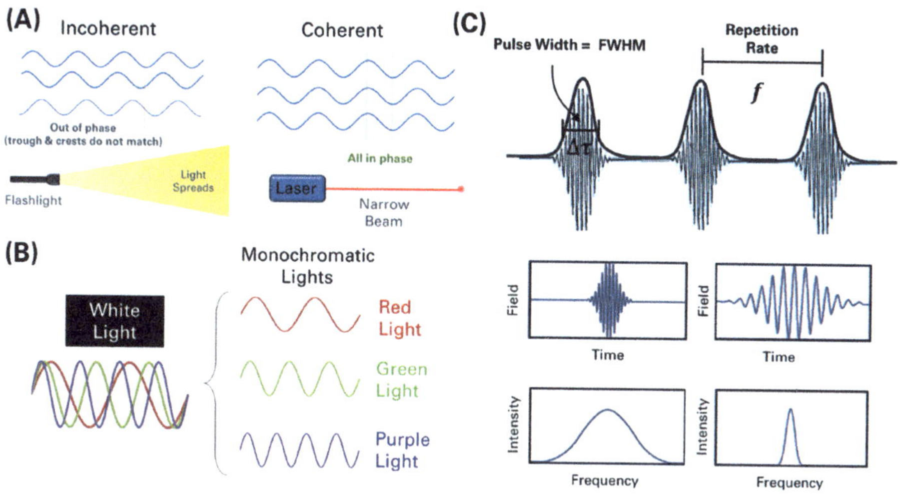

Figure 4.8: Comparison of (A) incoherent and coherent light, and (B) monochromatic versus white light. (C) Depiction of the concepts of pulse width and repetition rate in a pulsed laser.

Another advantage of lasers is that they can produce highly polarized light. Optical components in the laser cavity such as Brewster windows or mirrors allow selective production of linearly polarized light. In certain cases, polarizers or waveplates are used to select a specific polarization state. In certain lasers, the gain medium with anisotropic crystals or semiconductors have polarization-dependent properties. In this case, photons are generated with a specific polarization state influenced by the incoming pump photons. Stimulated emission ensures that emitted photons have the same polarization, helping to maintain a stable polarization state. Laser outputs are often monochromatic. Sunlight or light from fluorescent bulbs contains most of the colors in the visible range. The mechanisms of laser operation described above produce monochromatic light consisting of a single wavelength or more accurately, a very narrow band of wavelengths. In addition to stimulated emission, the electronic transition between specific energy levels within the lasing medium determines the specific wavelength of the laser beam. Monochromaticity is very important as it allows tight focusing with reduced chromatic aberration (e.g., blurring and distortion of image). Use of monochromatic light is required for sharper images and accurate measurements as in achieving high spectral resolution.

There are several advantages of a pulsed laser over a continuous wave laser. For instance, they yield high peak power in short bursts. This allows time for cooling in between pulses and therefore pulsed lasers offer reduced thermal damage. The high peak intensity increases the probability of nonlinear interaction of light with matter (see Chapter 6). Figure 4.8(C) highlights some of the key features of a pulsed laser. Each pulse in the pulse train is described as a wave packet – representing the localized electric field burst in the time domain. The envelope (such as the Gaussian function shown) describes the intensity amplitude of the wave packet. As can be seen, pulses are separated by periods of no light emission. This time between pulses is known as the repetition rate (f_{RR}). For many types of lasers, especially mode-locked lasers, the repetition rate is determined by the cavity length. It is the inverse of the time it takes for a pulse to complete a round trip within the cavity. The full width at half maximum of the pulse profile is defined as the pulse duration or pulse width, $\Delta\tau$. The energy within each pulse all varies depending on the specific laser. In practice, one can measure the average power of the laser, P_{avg}. From this measurement, the energy per pulse can be determined as follows:

$$E_{pulse} = \frac{P_{avg}}{f_{RR}} \tag{4.17}$$

The peak power, P_{peak} is calculated as

$$P_{peak} = \frac{E_{pulse}}{\tau} \tag{4.18}$$

As an example, a laser with an average power of 500 mW, 80 MHz repetition rate, and a pulse duration of 100 fs produces a peak power of 62.5 kW. We can go further and divide this number by the laser beam spot area to obtain a quantity called peak intensity (I_{peak}):

$$I_{peak} = \frac{P_{peak}}{\pi w_f^2} \tag{4.19}$$

For a focused Gaussian beam, the radius of the spot area is related to the focal length (f), the beam waist radius (w_0), and the wavelength (λ) of the light and can be estimated as follows: $w_f = \frac{\lambda f}{\pi w_0}$. Considering a focused beam spot area of 500 , we get a peak intensity of 1.25×10^{14} W/m^2! For comparison, the peak intensity of sunlight on earth on a clear day when the sun is at its apex is approximately 1,000 W/m^2.

In pulsed lasers, the pulse width is inversely related to its spectral bandwidth ($\Delta\nu$). Ultrafast lasers, which produce extremely short pulses, inherently have a broad bandwidth. For a Gaussian laser beam, $\Delta\tau \cdot \Delta\nu = 0.441$. Broadband lasers with wide range of emitted wavelengths provide versatile applications. By adjusting parameters like pulse width, repetition rate, and energy, pulsed lasers allow fine-tuning of the laser and its interaction with the material. Additionally, the ultrafast laser provides time resolution to measure fast molecular events that occur on extremely short timescales. A technique called pump-probe spectroscopy, which utilizes short-pulsed lasers, allows the study of ultrafast dynamics in materials. Figure 4.9 illustrates the concept of pump-probe spectroscopy. In brief, two laser pulses are used – a pump pulse and a probe pulse. The former excites the sample, and the time-delayed probe pulse monitors the changes induced by the pump. The time delay between these pulses is controlled mechanically using a delay stage. The response of the material system is plotted against the time-delay, which maps out the dynamic profile. The time-resolution is dictated by the pulse-width of the laser.

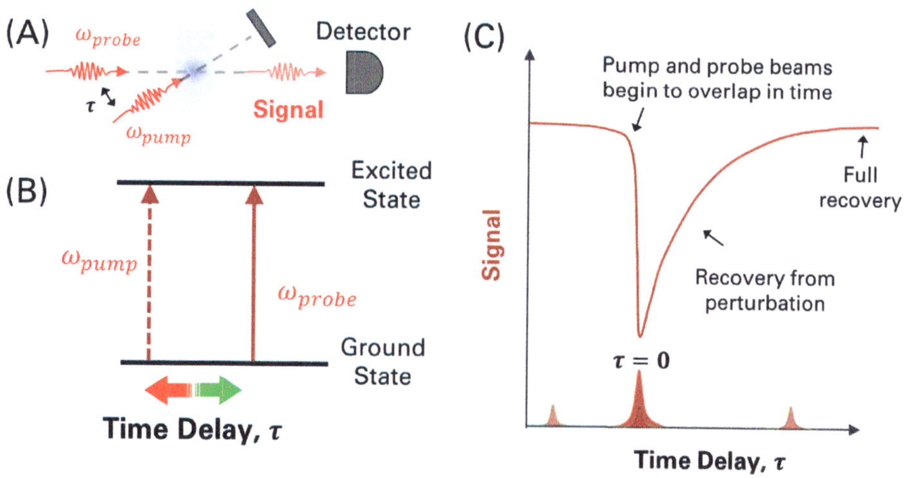

Figure 4.9: Pump-probe spectroscopy illustrated: (A) experimental overlap of the pump and the probe beam, with a delayed arrival time, at the sample. (B) Energetically, both pulses have same frequency, but with a time delay between them. (C) Schematic of an ultrafast dynamics trace based on the pump-probe method. At $\tau = 0$, the pump and the probe beams overlap completely, and the signal drops in a process known as bleaching.

Lasers offer significant advantages over conventional light sources, leading to their extensive use across both applied and fundamental research. The following chapters will highlight key applications of lasers in investigating the fundamental properties of colloids and their interfaces.

End-of-Chapter Problems

1 Concepts and definitions:
   (A) List and briefly describe the major nonlaser-based techniques that are common for colloidal and surface analysis.
   (B) What is the difference between spectroscopy and spectrometry?
   (C) Define the terms: linear polarization, circular polarization, elliptical polarization, electric field, intensity, oscillator strength, coherence.
   (D) Explain the type of transitions and physical insights that can be gained from the following spectroscopic techniques: microwave, infrared, UV-Vis.
   (E) How is light (laser) used to study colloidal suspension?
   (F) Describe the following processes: spontaneous emission, stimulated emission, population inversion, and lasing.
   (G) Highlight the differences between elastic and inelastic light scattering. Provide examples.
   (H) Explain the difference between absorbance and turbidity of a dispersion.
   (I) Discuss the three major scattering theories (Rayleigh, RGD, and Mie) discussed in this chapter. Describe their limitations and applications in colloidal science in the context of size parameter.
   (J) What factors influence light-scattering intensity and direction of the scattered light?
   (K) Explain Raman spectroscopy and how it complements IR spectroscopy. Compare Stokes versus anti-Stokes Raman shifts.
   (L) What are the advantages of a laser light source over ordinary light source? What is the difference between a continuous-wave and a pulsed laser?
   (M) Describe pump-probe spectroscopy.

2 Asymmetrical flow field-flow fractionation (AF4) coupled with multiangle laser light scattering (MALLS) and a concentration detector (e.g., UV or differential refractive index) is used to characterize a sample of colloids. The system has a channel thickness of 350 μm. The flow rates are set at 0.5 mL/min and a constant crossflow for separation.
   (A) Explain the fundamental principle by which AF4 separates the nanoparticles. What role does the laser detector (MALLS) play in characterizing the size and potentially the shape of the separated fractions?
   (B) A sample containing two distinct populations of spherical polystyrene beads – 50 and 100 nm in diameter – are separated. Which population will elute first from the AF4 channel? Explain.
   (C) Another technique that uses light scattering is nanoparticle tracking analysis (NTA). Describe the principle of NTA. What physical property of the nano-

particles does NTA directly measure using the laser and camera system to determine particle size?

3 For a material with a frequency-dependent susceptibility, it is found that the oscillator strength parameter, A, is 2.5×10^{32} m^{-3}s^{-2}, the resonance frequency is $\omega_0 = 5 \times 10^{15}$ rad · s^{-1}, and the damping constant, γ, is 1×10^{14} rad · s^{-1}. Calculate the complex susceptibility $\chi^{(1)}$, complex permittivity ε, and the complex refractive index, n at $\omega = 4 \times 10^{15}$ rad · s^{-1}. Recall, $\varepsilon = n^2 = 1 + 4\pi\chi^{(1)}$ and the linear susceptibility (eq. (4.10)) can be expressed as $\chi^{(1)} = \frac{A}{\omega_0^2 - \omega^2 - i\omega\gamma}$.

4 Let us consider light scattering from silica particles of radius 5 nm dispersed in an aqueous medium.
- (A) Does the particle size meet the criteria for Rayleigh scattering for both blue ($\lambda = 450$ nm) and red light ($\lambda = 680$ nm)?
- (B) By what factor is the intensity of the scattered blue light greater than the intensity of the scattered red light at a 90° angle to the incident beam?
- (C) What color is the suspension likely to appear when viewed from the side and directly through?

5 A researcher is using surface-enhanced Raman spectroscopy (SERS) with a colloidal gold nanoparticle solution to identify an unknown organic molecule. An excitation laser with a wavelength of 632.8 nm is used. The Raman spectrum collected shows a strong Stokes line peak for the unknown molecule at an absolute scattered wavelength of 675.1 nm. For the SERS measurement, 1.0 μM of the unknown compound was mixed with colloidal gold and SERS intensity of 85,000 counts per second. The number of molecules absorbed at the gold particles that are effectively probed by the laser is 1.2×10^7 molecules.
- (A) Calculate the Raman shift in cm^{-1} of this specific vibrational mode.
- (B) If the standard Raman measurement of a 1.0 mM solution of the unknown compound yields an intensity of 400 counts for the same peak, calculate the apparent surface enhancement factor. The number of molecules effectively probed by the laser in this standard setup is 5.0×10^{11} molecules.
- (C) In the SERS experiment, the total available surface area of the gold particles is 1.8×10^7nm^2. Calculate the surface coverage of the unknown compound.

6 Consider an experiment to measure the concentration of uniform, spherical aerosol particles in a chamber using a laser beam. The primary mechanism by which the laser light is attenuated is scattering (turbidity). Answer the following question based on this information:

Laser wavelength	Pathlength	Incident light intensity	Molecular scattering cross section	Measured transmitted light intensity
632.8 nm	10 cm	1.0 W/m^2	$1.2\times10^{-27}\,\frac{m^2}{molec}$	0.45 W/m^2

(A) Calculate the turbidity of the medium.

(B) Using the calculated turbidity and the provided molecular cross-section, determine the number concentration of the aerosol particles in units of particles per m^3.

(C) Calculate the total number of particles within the illuminated volume, assuming the laser beam has a cross-sectional area of 5.0×10^{-6}m^2.

7 Laser characterizations:

(A) A He-Ne laser emits light at a center wavelength 632.8 nm with a very narrow spectral linewidth 0.01 nm. An incandescent light bulb emits light with a center wavelength of approximately 600 nm and a broader spectral linewidth of around 100 nm. Calculate the coherence length for both light sources and explain the difference in their coherence properties.

(B) Find the peak power and peak intensity for both a 100 W CW laser and a 100 W pulsed laser, with the pulsed laser having repetition rate of 10 kHz and 10.0 ns pulse duration. The beam radius for both lasers is 1.0 mm.

(C) What is the effect of pulsed width, beam spot size, and the focal length of the focusing lens on the peak intensity of a pulsed laser?

8 A common laser beam is often linearly polarized. When linearly polarized light passes through a polarizing filter (polarizer), the transmitted intensity is given by Malus's law: $I = I_0\cos^2(\theta)$ where I_0 is the incident intensity and θ is the angle between the incident polarization direction and the polarizer's transmission axis. A half-waveplate (HWP) introduces a phase shift of π between the electric field components parallel to its fast and slow axes. A HWP rotates the plane of linear polarization by an angle twice the angle between its fast axis and the incident polarization plane. A quarter-waveplate introduces a $\pi/2$ phase shift between the electric field components along its fast and slow axes.

(A) A laser beam is horizontally polarized. A HWP is inserted into the beam path with its fast axis oriented at 30° relative to the horizontal polarization plane. What is the polarization state of the light emerging from the HWP?

(B) A horizontally polarized laser beam with a power of 100 mW (milliwatts) passes through a polarizing filter. The transmission axis of the filter is oriented at an angle of 30° from the horizontal. What is the power of the laser beam after it emerges from the filter?

(C) A vertically polarized laser beam with an electric field amplitude E_0 is directed at a quarter-wave plate. How should the quarter-waveplate be oriented to produce circularly polarized light? Describe the resulting electric field components.

(D) You have an unknown waveplate and a linear polarizer. You place the waveplate after a vertically polarized laser source. When you rotate the polarizer after the wave plate, the intensity of the transmitted light varies, but never reaches zero (extinction). What type of waveplate could it be, and what is the resulting polarization? You rotate the wave plate itself (around the beam axis) and find an orientation where the polarizer can achieve full extinction. After finding this orientation, you rotate the wave plate by exactly 45°. Now, the intensity reading on the detector remains constant regardless of how you rotate the analyzer. What type of wave plate is it, and what is the output polarization in this configuration?

5 Application of Light Scattering and Colloidal Properties

We have seen that the intensity, angular distribution, and polarization of the light scattered from colloids can depend on various colloidal properties. Light scattering techniques are thus useful to study the size, charge properties, and molecular weight of colloids and macromolecules. Elastic light scattering methods can be broadly divided into two categories: static light scattering and dynamic light scattering, abbreviated as SLS and DLS, respectively. Both use a monochromatic and coherent light source from a laser and analyze the scattered light from a colloidal suspension. Another common light scattering technique that facilitates the determination of zeta potential is called electrophoretic light scattering (ELS). In this method, the frequency shift of scattered laser light is detected, which is used to calculate the velocity of charged particles under an applied electric field, i.e., their electrophoretic mobility. Many modern instruments can measure SLS, DLS, and ELS using a single compact unit. In this chapter, the basics of these laser light scattering measurements are covered. Experiments and data analysis based on these methods in deriving colloidal properties are presented. At the end of this chapter, applications of lasers in other areas involving colloidal science, such as the synthesis of nanomaterials, particle manipulation, and remote sensing, are summarized.

5.1 Principles of Light-Scattering Techniques

Rayleigh, RGD, and Mie theories explain various features of the scattered light. Here, the focus is on how the scattered intensity is measured and processed to acquire information about colloids. In static or classical light scattering, the intensity of the scattered light detected is *time-averaged*. The molecular weight of the macromolecules or particles and the radius of gyration can be obtained using SLS. The principle of SLS is the basis of other techniques such as multiangle (laser) light scattering (MALLS or MALS for short), low-angle light scattering, right-angle light scattering, and laser diffraction for large particles. In the 1950s, prior to the invention of lasers, MALS used collimated light from conventional light sources such as mercury vapor lamps. Subsequently, lasers served as an integral light source.

DLS works by measuring *intensity fluctuations* that occur over time due to Brownian motion (Chapter 2). As a result, DLS allows determination of particle diffusion coefficient, which is related to its hydrodynamic radius (R_h) according to the Stokes-Einstein equation (2.14). In other words, particle (Brownian) motion is averaged out in SLS, whereas in DLS, the effect of particle diffusion is measured. Other common names for DLS are photon correlation spectroscopy and quasi-elastic light scattering. While the science of light scattering and Brownian motion was already known, the

 | https://doi.org/10.1515/9783111018386-005

invention of DLS instruments did not blossom until the late 1960s and early 1970s, after the inventions of the laser and the digital autocorrelator. The latter is a critical component of intensity fluctuations analysis.

ELS, also known as laser Doppler electrophoresis, detects the Doppler shift in the scattered light. The frequency of a wave, such as a sound or light wave, shifts when its source or the observer is moving relative to each other. This is known as the Doppler effect, named after Christian Doppler, an Austrian physicist. The Doppler shift is proportional to the particle's velocity experiencing electrophoresis, which in turn can be related to zeta potential (Chapter 3). Since the particle velocity is measured, laser Doppler velocimetry is another term used for this approach. ELS uses the principle of laser Doppler velocimetry and allows the acquisition of information about the charge properties of colloids. A related technique that uses the same principle but is more sensitive is called phase analysis light scattering (PALS). This method measures the phase shift instead of the frequency shift of the scattered light as a function of time.

The basic physics of light scattering is shown in Figure 5.1. The laser beam illuminates suspended particles, and scattered photons are detected. A key parameter is the scattering wavevector, $\vec{q}$. It is the difference between the wavevector of the incoming laser beam and the scattered wave. For negligible dispersion, the relationship between $\vec{q}$ and the scattering angle (θ) is as follows:

$$|\vec{q}| = \frac{4\pi n_0}{\lambda_0}\sin\left(\frac{\theta}{2}\right) \tag{5.1}$$

Scattering wavevector represents the momentum transfer of the light-particle interaction and its magnitude depends on the fundamental wavelength (λ_0), and refractive index (n_0) of the dispersant at λ_0. The wavevector plays a critical role in all light scattering processes.

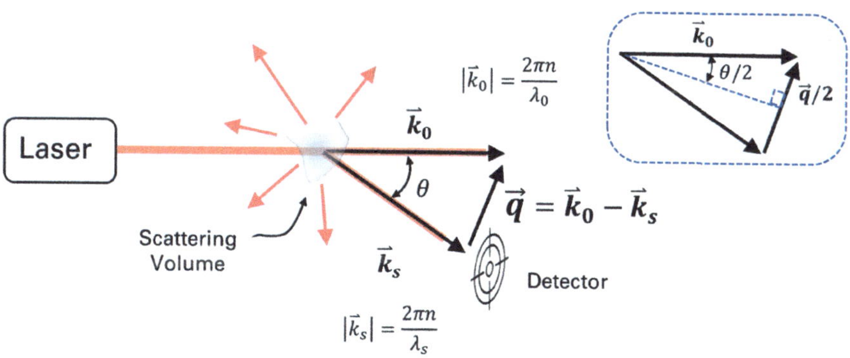

Figure 5.1: Scattering geometry and definition of the scattering wavevector.

5.2 Dynamic Light Scattering

Consider a monochromatic light from a laser source impinged upon a collection of stationary particles as shown in Figure 5.2(A). If we detect the scattered light on a screen at a certain fixed distance, we will see a speckled pattern composed of dark and bright spots. This is because the scattered light undergoes constructive and destructive interferences before reaching the screen or the detector. Figure 5.2(A) demonstrates this phenomenon. Certain fields are in phase, and others are out-of-phase. If the scattered light intensity from the stationary particles is monitored over time, it will appear constant, barring any noise associated with the measurement. However, in a colloidal system, all particles are moving. In Figure 5.2(A), right panel, a simplified picture is depicted – only one particle is shown to have moved while the other two are stationary. Yet, we see that the location where the constructive and destructive pattern occurs has swapped. Because of Brownian motion, the intensity of the scattered light fluctuates. The scattered light randomly alternates between dark and bright spots.

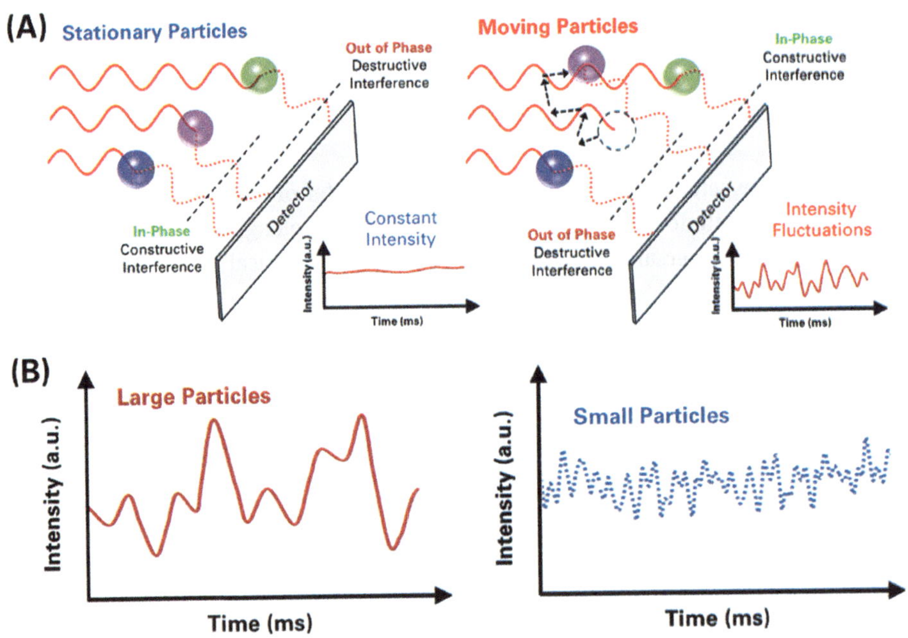

Figure 5.2: (A) A cartoon illustrating constructive and destructive interference of scattered light from stationary and moving particles. (B) Comparison of intensity fluctuations in large versus small particles.

In DLS, a laser is incident on a colloidal suspension, and the scattered light is detected at an angle. Typically, a photomultiplier tube or an avalanche photodiode is used. Large particles diffuse slower than small particles (eq. 2.14). Thus, the variation in intensity (Figure 5.2(B)) is relatively gradual for larger particles compared to that of the smaller ones. The incessant Brownian motion causes a slight frequency shift (Γ) of the incident monochromatic light, the extent of which is dependent on the particle diffusion coefficient, D, as follows: $\Gamma = Dq^2$, where q is the scattering wavevector. It is therefore evident that the intensity fluctuation contains information about the particle size. The timescale at which the fluctuation in these intensities occurs must be quantified to gain a quantitative understanding of the size of the suspended particles. Fluctuations occur on the timescale where particles move about one wavelength of light. For a 1-µm polystyrene particle excited by 632 nm light, this timescale (eq. (2.13a)) is approximately 900 ms. A signal processing method called autocorrelation is employed in the DLS technique. It measures the similarity of a signal with a delayed version of itself and maps out the extent of similarity over a period of time, typically ns to μs. Along with a laser, the autocorrelation technique is an integral part of DLS. A laser is important as it provides a coherent, monochromatic, light source of specific polarization.

In DLS experiments, an intensity correlation function (second-order correlation function), $g^2(\tau)$, is measured. To understand this quantity, let us refer to Figure 5.3(A). The scattered intensity ($I(t)$) at four different time intervals or delay time, τ, is shown. In a typical DLS experiment, τ can range from a few tens of nanoseconds to a few hundreds of milliseconds. Pictorially, we see that if τ is close to zero, the two intensity profiles overlap the most, i.e., they are correlated. As τ approaches infinity, the intensity profiles do not match, and the signals are said to be completely uncorrelated. In between $\tau = 0$ and $\tau = \infty$, the correlation decreases with increasing τ. This concept can be written mathematically as

$$g^2(\tau) = \frac{\langle I(t) \cdot I(t+\tau) \rangle}{\left\langle I(t)^2 \right\rangle} \tag{5.2}$$

In this equation, $I(t)$ represents the intensity at an arbitrary time t, and $I(t+\tau)$ is the intensity after the delay. The angular brackets, $\langle\ \rangle$, denote an average quantity. Division by $\langle I(t) \rangle^2$ makes the intensity correlation functions a dimensionless normalized quantity. In a typical light scattering experiment, intensity fluctuations measured at two different times t and $t+\tau$ appear as statistical noise. If τ is sufficiently smaller than the characteristic fluctuation time, these intensities measured in the time interval τ are correlated.

As an example, $\langle I(t) \cdot I(t+\tau) \rangle$ is $\langle I^2 \rangle$. The quantity $\left\langle I^2 \right\rangle$ is either greater than or equal to $\langle I \rangle^2$. Thus, for a perfectly correlated signal, $g^2(\tau = 0) > 1$. When $\tau \to \infty$, the numerator approaches $\langle I \rangle^2$ and $g^2(\tau = \infty)$, a value of 1. At $\tau = \infty$, $I(t)$ and $I(t+\tau)$ become statistically independent. For independent random variables $\langle A \rangle$ and $\langle B \rangle$, the expectation of their product is their product of expectations, i.e., $\langle AB \rangle = \langle A \rangle \langle B \rangle$. Our discussion thus far has been pertinent to intensity correlation function, $g^2(\tau)$. However, scattered light from particle movement is correlated by means of the electric field cor-

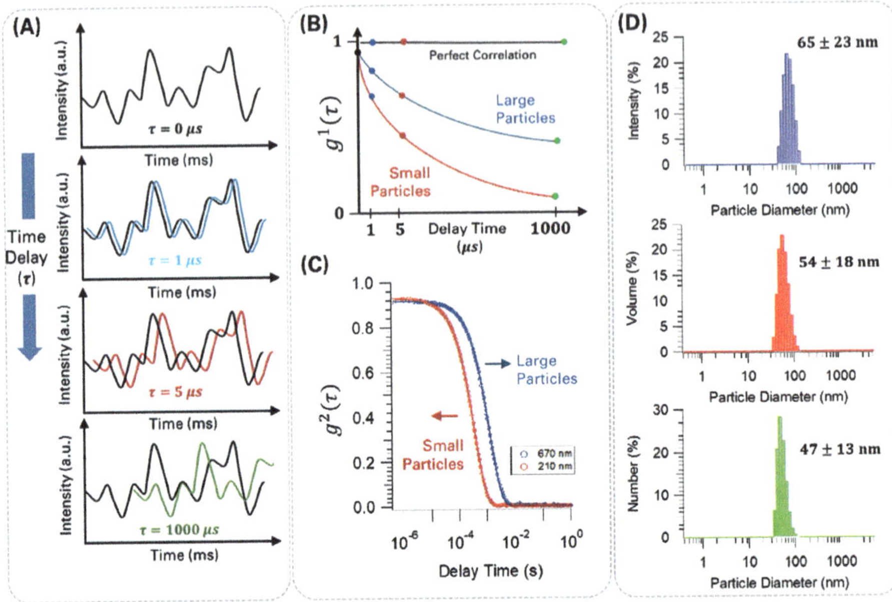

Figure 5.3: (A) A depiction of intensity fluctuation and relative correlation. (B) Exponential decay of $g^1(\tau)$ and the effect of particle size. (C) Experimental data: intensity-intensity correlation function versus delay time (log-scale), also known as correlogram, for two different latex particles. The solid traces are fit of experimental data using eq. (5.4). (D) Comparison of intensity, volume, and number distribution histograms.

relation function, $g^1(\tau)$. DLS involves homodyne detection; only the scattered light is detected and not the incident beam. If the photon counting of the scattered light by the detector is a random Gaussian process, the Seigert relation connects the intensity and the field correlation functions as follows:

$$g^1(\tau) = \sqrt{g^2(\tau) - 1} = e^{-Dq^2\tau} \tag{5.3}$$

For a suspension of monodispersed spherical particles, $g^1(\tau)$ decays exponentially with the delay time. Figure 5.3(B) shows the effect of particle size. A perfect correlation would be observed for stationary particles. Rearranging eq. (5.3), a general expression for the intensity-intensity correlation function can be written:

$$g^2(\tau) = A + \beta e^{-2\Gamma\tau} \tag{5.4}$$

where the decay rate is $\Gamma = Dq^2$. In this equation, β is the coherence factor, which depends on the scattering properties of the colloids, the optical geometry, and the detector area. The parameter A represents the baseline and can be normalized to 1. The measured decay is the intensity-weighted sum of the decay of the individual particles.

Particles of different sizes exhibit different decay rates. Equation (5.4), which contains D as a fitting parameter, is used to fit experimental data as shown in Figure 5.3(C). The scattering vector q is a known parameter (eq. (5.1)). Once the diffusion coefficient is determined, the size of the particle is calculated using eq. (2.14). For this calculation, the viscosity and refractive index of the dispersant must be known. The particle radius calculated based on this technique is called the hydrodynamic radius, R_h. It is not the exact radius of the hard sphere but reflects the particle's apparent size as it moves through a fluid. In an aqueous dispersant, the term "hydrodynamic" emphasizes the fact that the particle includes a hydration shell around its surface.

A common laser wavelength in modern DLS instruments is 632.8 nm (HeNe laser), as it is compact and relatively inexpensive. The angle of detection is typically set at 90° (side scatter), 173°, or 175° (back scatter), and forward scatter around 13° to 15°. Figure 5.3(C) shows experimental data and the influence of particle size on the correlation function. Back scattering is useful for highly concentrated samples as it minimizes the possibility of multiple scattering reaching the detector. A 90° detection is not ideal for turbid samples but effective for small particles, as large particles mostly scatter in the forward direction (see Chapter 4). In multiangle dynamic light scattering, correlation data from several detection angles are combined. Knowledge of Mie theory is incorporated to generate a higher resolution particle size distribution. The volume of the sample can vary, ranging from a few microliters in a specialized setup to 1–2 mL for standard cuvettes. The typical laser beam focus is about 0.2 mm. Most commercially available instruments include software and perform the fitting automatically. The algorithms extract the decay rate and generate an intensity distribution plot where the x-axis shows a distribution of size classes and the y-axis, the relative intensity of the scattered light, as shown in Figure 5.3(D), top panel.

The intensity-weighted average diffusion coefficient $\langle D \rangle$ is obtained from DLS experiments is based on eq. (5.5), which describes its relationship to the mass (M_i), the diffusion coefficient (D_i), and the number (N_i) of the ith particle. The form factor $P(q, r)$ depends on the particle size, the scattering vector (eq. (5.1)), and the refractive index of the particles. An example of the form factor has been presented in Section 4.4.1. If the particles are much smaller than the wavelength of the incident light, $P(q, r)$ can be approximated to be 1. The denominator is the total scattered light intensity from all the particles, and the numerator weights the contribution of each diffusion coefficient:

$$\langle D \rangle = \frac{\sum_i N_i M_i^2 P(q,r) D_i}{\sum_i N_i M_i P(q,r)} \tag{5.5}$$

5.2.1 Monomodal Versus Nonmonomodal Distribution

To obtain reliable information about the particle size distribution from the correlogram, there are two major approaches of data analysis – monomodal distribution versus the nonmonomodal distribution methods. Equation (5.4), which shows a single exponential decay, is applicable for colloidal suspensions of uniform size – a monomodal distribution where the particles have a narrow size distribution around a mean. The correlation function $g^1(\tau)$ can be expressed as an integral weighted by the distribution function:

$$g^1(q,\tau) = \int_0^\infty G(\Gamma)e^{-\Gamma\tau}d\Gamma \tag{5.6}$$

Here, $G(\Gamma)$ denotes the relative intensity of light scattered with a decay constant Γ. This method of analyzing the data is known as the cumulant analysis method, which has been accepted and recommended by the International Standards Organization. It uses Taylor series expressions on the natural logarithm of the integral. Accordingly, the following expression is obtained:

$$\ln\left(g^2(\tau) - A\right) = \ln\beta - \bar{\Gamma}\tau + \frac{k_2}{2!}\tau^2 - \frac{k_3}{3!}\tau^3 + \frac{k_4}{4!}\tau^4 + \cdots$$

where $\bar{\Gamma}$ is the mean decay rates, or the cumulants mean, assuming a Gaussian distribution. This leads to the intensity weighted average mean size or a parameter called Z-average (Z_{avg}):

$$Z_{avg} = \frac{1}{\bar{\Gamma}}\frac{k_B T}{3\pi\eta}\left(\frac{4\pi n_0}{\lambda_0}\sin\left(\frac{\theta}{2}\right)\right)^2 \tag{5.7}$$

The parameters k_2, k_3, and k_4 represent the variance, skewness, and kurtosis of the measured distribution, respectively. The key advantage of the approach is that the experimental data $\ln\left(g^2(\tau) - A\right)$ can be fit using a polynomial function and the fit parameters extracted. The key parameters of interest are the mean value of the size and the width parameter, known as the polydispersity index (PDI). The latter is mathematically defined as $\frac{2k_2}{\bar{\Gamma}^2}$. Generally, higher-order cumulants are not obtained to prevent overfitting of the data. The implication is that Z_{avg} is likely to be misinterpreted if the size distribution is very broad in the case of colloids exhibiting large dispersity. A low PDI value (typically < 0.05) indicates particles of uniform size distribution. PDI larger than 0.7 is recognized as having large dispersity and not suitable for quantitative analysis.

For analyzing colloidal size distributions of broad monomodal or multimodal nature, there are several data analysis methods available. These include nonnegatively constrained least squares, exponential sampling, and the constrained regularization method for inverting data (CONTIN) techniques. Nonmodal analysis does not assume a specific type (e.g., Gaussian) of distribution and is therefore applicable for colloids that exhibit large dispersity. In the nonnegatively constrained least square method, a discrete number of decay constants Γ_i representing the decay rates $G(\Gamma)$ spaced in a linear or logarithmic pattern over a chosen ranges of Γ is selected. Laplace transform of eq. (5.4) is used, and for Γ_i, coefficients b_i are obtained from the best fit of the data with the constraint of nonnegative size distribution. These coefficients are obtained by the minimization of the following expression:

$$\chi^2 = \sum_j^N \left[g^1(\tau_j) - \sum_{i=1}^M b_i e^{-\Gamma_i \tau_j} \right]^2$$

The parameters N and M are the number of data points in the fit and the number of decay constants in the distribution, respectively. The CONTIN method analyzes the autocorrelation function using an inverse Laplace transformation and regularization. It utilizes a nonlinear statistical technique to smooth the solution, and the algorithm minimizes the regularized residual. These approaches provide better analysis of broad particle distributions. Most modern DLS instruments are equipped with software that can perform these types of data analyses.

5.2.2 Intensity, Number, Versus Volume Distributions

The analysis described above results in intensity-weighted size distributions. The intensity distribution reflects the light scattered by particles. Based on Rayleigh scattering (eq. (4.11)), the scattered intensity scales with the sixth power of the particle's diameter. The relationship between scattered intensity and particle size is not a simple power-law for Mie scattering. It oscillates with particle size, especially as the particle size becomes larger relative to the wavelength of the excitation light. Nonetheless, intensity distribution is often heavily influenced by larger particles compared to the smaller particles. A raw intensity distribution profile shows larger relative peaks for larger particles, even if they are present in relatively small amounts. For this reason, and to better understand the actual proportion of colloids, volume-distribution may be needed. The intensity distribution profile can be converted to a volume-weighted or number-weighted distribution using Mie theory. This requires knowledge of the sample's optical properties, such as the refractive index of the dispersant and the particles. The transformation also assumes that the particles are spherical. Mathematically, intensity is divided by a factor related to the particle volume. Number distribution is obtained upon further correction of the volume-weighted distribution. Software packages available in modern instruments provide these calculations.

The following definitions describe the scaling factors of the number ($\%N_i$), volume ($\%V_i$), and intensity ($\%I_i$, assuming Rayleigh scattering) distribution with respect to particle size, d_i:

$$\%N_i = \frac{100N_i}{\sum_i N_i},\ \%V_i = \frac{100N_i d_i^3}{\sum_i N_i d_i^3},\ \%I_i = \frac{100N_i d_i^6}{\sum_i N_i d_i^6}$$

Comparing these equations, it is evident that the intensity distribution is skewed toward larger particles, with the number distribution most biased toward smaller sizes, as is apparent in Figure 5.3(D). Particle diameter obtained from intensity distribution is the most fundamental and is always correct, as it is the immediate output of the DLS technique. It bears the least number of assumptions. It is also best for identifying very small numbers of aggregates or dust particles in a colloidal suspension, as larger objects create a substantial signal. Volume-weighted diameter is best compared with results that are mass or concentration based, whereas number-weighted size compares well with microscopy-based techniques. DLS is applicable to particle sizes in the range of 1 nm to 10 μm. Brownian motion becomes less pronounced as particle size increases above the colloidal range. For larger particles, SLS or laser diffraction techniques are suitable.

5.3 Static Light Scattering

The term "static" in SLS refers to the fact that scattered intensity is averaged over time. SLS can be used to measure the size and shape factors of particles, but it is commonly used to determine the molecular weight of macromolecules. When size analysis is performed, SLS is often synonymous with the laser diffraction method. A laser is used to illuminate a colloidal sample contained in a cuvette. One or many detectors are used to collect the scattered light intensity ($I(\theta)$) as a function of scattering angle θ. As we have seen in Chapter 4, $I(\theta)$ is dependent on particle size, relative refractive index, and the wavelength of the light. When the particle size is small compared to the wavelength ($< \frac{\lambda}{20}$), Rayleigh theory can be applied suitably to determine molecular weight (M_w).

Referring to eq. (4.12), the scattered intensity is proportional to molecular polarizability, α. In a suspension, α fluctuates as the concentration of the solute and the density of the solvent fluctuate within a small volume element. Thus, it is the fluctuation, $\langle \Delta\alpha^2 \rangle$, that is relevant to eq. (4.12). The density fluctuation can be empirically subtracted from scattering measurements of the solvent alone. As for the concentration fluctuation ($\langle \Delta c^2 \rangle$), it can be shown that $\langle \Delta\alpha^2 \rangle \propto n^2 \left(\frac{\partial n}{\partial c}\right)^2 \langle \Delta c^2 \rangle$, where n is the refractive index, c represents the concentration of the macromolecule, and $\frac{\partial n}{\partial c}$ is the refractive index increment. As an example, most unmodified proteins in aqueous solution have an approximate $\frac{\partial n}{\partial c}$ value of 0.19 mL/g. Based on fluctuation theory and thermodynamic arguments [57, 80], the following relationship is obtained, which is applicable for dilute solutions:

$$\frac{I_s}{I_0} = \frac{2\pi^2 n^2}{N_A \lambda^4 R^2} \left(\frac{\partial n}{\partial c}\right)^2 c\left(1+\cos^2\theta\right) \frac{1}{\frac{1}{M}+2Bc} \tag{5.8}$$

where B (also denoted as A_2) is the second virial coefficient relating osmotic pressure to concentration. It is a measure of the interaction between the dispersed particles. If B is positive, interparticle forces are repulsive, and if negative, the forces are attractive. $B=0$ corresponds to an ideal dispersion with no net interactions.

We can further define the Rayleigh ratio (R_θ) as follows:

$$R_\theta = \frac{I_s R^2}{I_0\left(1+\cos^2\theta\right)} \tag{5.9}$$

The Rayleigh ratio is essentially the quantity turbidity, τ, (Chapter 4), and has a unit of per length. Equation (5.8) expressed in terms of the Rayleigh ratio is

$$\frac{Kc}{R_\theta} = \frac{1}{M} + 2Bc \tag{5.10}$$

where $K = \frac{2\pi^2 n^2}{N_A \lambda^4 R^2}\left(\frac{\partial n}{\partial c}\right)^2$. All these optical parameters are wavelength dependent. A plot of $\frac{Kc}{R_\theta}$ (eq. (5.9)) as a function of macromolecule concentration under isothermal conditions is thus expected to yield a straight line with a slope and an intercept of $2B$ and $\frac{1}{M}$, respectively. This is the basis of molecular weight determination based on SLS. Note, eq. (5.10) is applicable for particles much smaller than the wavelength of the light. Moreover, it does not apply to charged macromolecules because interparticle electrostatic repulsions cause significant deviations from the ideal scattering behavior of neutral particles. For charged particles, measurements can be made in a suspension with added electrolytes to screen the charges.

5.3.1 Zimm-Debye Equation and Zimm Plot

An extension of eq. (5.10) that applies to larger particles is based on the Debye scattering theory. The assumptions that particles are smaller than λ and that they have a refractive index that is not too large still hold. The modified equation (eq. (5.11)) now includes a form factor, $P(\theta)$. The Rayleigh ratio has also been replaced by ΔR_θ since the difference between the solution and pure solvent is involved in the experimental measurements:

$$\frac{Kc}{\Delta R_\theta} = \frac{1}{P(\theta)}\left(\frac{1}{M} + 2Bc\right) \tag{5.11}$$

This is known as the Zimm-Debye equation. In the limit of small particles, less than $\frac{\lambda}{20}$, $P(\theta)=1$. A general form factor based on RGD theory has been introduced in eq. (4.15).

It contains information about the internal structure of the particles. If qr is small, the form factor can be further approximated as

$$\frac{1}{P(\theta)} = 1 + \frac{16\pi^2 n^2 \langle R_g^2 \rangle}{3\lambda^2} \sin^2\left(\frac{\theta}{2}\right) \tag{5.12}$$

The quantity R_g is called the radius of gyration. It is a measure of how mass is distributed around a center of gravity or an axis of rotation. For a given material, a smaller value of R_g represents a compact structure. Thus, R_g provides a measure of the macromolecule's extension in space. For a sphere of radius r, $\langle R_g^2 \rangle = \frac{3r^2}{5}$. For a random coil and a thin rod, we have $\frac{r_{\mathrm{coil}}^2}{6}$, and $\frac{L^2}{12}$, respectively. In this case, r_{coil} is the end-to-end distance and L is the length of the rod molecule. Figure 5.4(A) shows how the form factor varies as a function of qr for different shapes of particles. In essence, the form factor is related to the size parameter of the macromolecule, and it can be determined based on SLS. Measurements often involve determining the ratio of $P(\theta)$ at two different angles. For example, the ratio $P(\theta = 45)/P(\theta = 135)$ is referred to as the dissymmetry ratio (Z). For isotropic scattering, when particles are significantly smaller than the wavelength of the light, $Z = 1$. The dissymmetry ratio provides a useful insight into the overall conformation of a macromolecule.

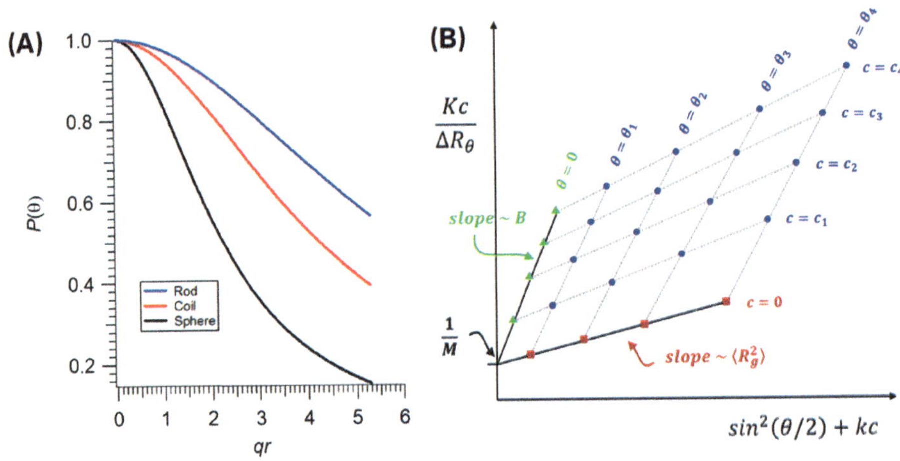

Figure 5.4: (A) Simulation of the form factor $P(\theta)$ as a function of qr for macromolecules of different shape. (B) Depiction of a Zimm plot with $\frac{Kc}{\Delta R_\theta}$ values at arbitrary detection angles and concentrations. The experimental data points are shown as blue circles. The extrapolated data points to zero concentration and $\theta = 0$ are shown as red squares and green triangles, respectively.

Let us now return to eq. (5.11). To determine M_w, SLS instruments are calibrated against a well-known solvent such as toluene or benzene. The Rayleigh ratio is measured at various angles for a fixed concentration and at various low concentrations at a fixed angle. A Zimm plot, which is a plot of $\frac{Kc}{\Delta R_\theta}$ versus $\sin^2\left(\frac{\theta}{2}\right) + kc$ is created. The parameter k is a scaling factor that brings the angle and concentration-dependent data points within the same numerical range. A representation of a Zimm plot is shown in Figure 5.4(B). Notice the slopes at different angles or concentrations are not necessarily the same. Thus, extrapolation to zero concentration and zero angle is critical. Extrapolation to zero concentration removes the effect of nonideal behavior due to solvent-macromolecule interactions. The effect of particle size is removed by extrapolation to $\theta = 0$. The y-intercepts of both extrapolated plots merge and give the molecular weight of the macromolecule. The slope of the angle-dependent data at zero concentration is $\frac{16\pi^2 n^2 \langle R_g^2 \rangle}{3\lambda^2 M}$ (eq. (5.12)). The slope of the concentration-dependent data at $\theta = 0$ is related to the second virial coefficient.

A Zimm plot is not always linear. In a nonideal suspension, the presence of large particles, higher concentrations, and molecular interactions can lead to deviations from eq. (5.11). If the solution is a mixture of different sizes of particles or macromolecules, the calculated molecular weight and radius of gyration will be an average value. As we have seen with DLS techniques, the weighting strongly favors the larger molecular weights. The average values are called the weight-average molecular weight and the z-average radius of gyration.

5.4 Electrophoretic Light Scattering

In Chapter 3, the concept of electrophoretic mobility, μ, has been introduced. It has been shown that μ is related to the zeta potential of a particle travelling under an applied field. The method of ELS allows measurement of μ. The basic mechanism involves the detection of a Doppler shift in scattered light from the particles moving under an applied electric field. Figure 5.5(A) provides a schematic diagram of the laser setup for an ELS experiment. It shows a heterodyne detection, in which a portion of the original unscattered (reference) laser beam with frequency ν_0 is combined (optically mixed) with the scattered light (sample). Figure 5.5(B) shows a typical microelectrochemical cell and the position where the light is impinged upon the sample. At this position, known as the stationary layer, electroosmotic flow (see Section 3.4) is zero. In a closed capillary system, this is the location where the flow moving in one direction is cancelled by a pressure-driven return flow moving in the opposite direction.

An oscillating mirror acts as a modulator that introduces a controlled frequency shift (ν_{mod}) to the reference laser beam. The overall frequency of the reference beam is $\nu_0 + \nu_{\mathrm{mod}}$. The scattered light also has a slight Doppler shift ($\Delta\nu$) due to particle motion and its overall frequency is $\nu_0 + \Delta\nu$. The interference between the reference and sample beams produces a "beat" frequency ($\nu_{\mathrm{mod}} + \Delta\nu$), which is the difference between the modulation frequency of the mirror and the Doppler shift from the particles. The modulator thus serves as an optical frequency mixer, transforming the

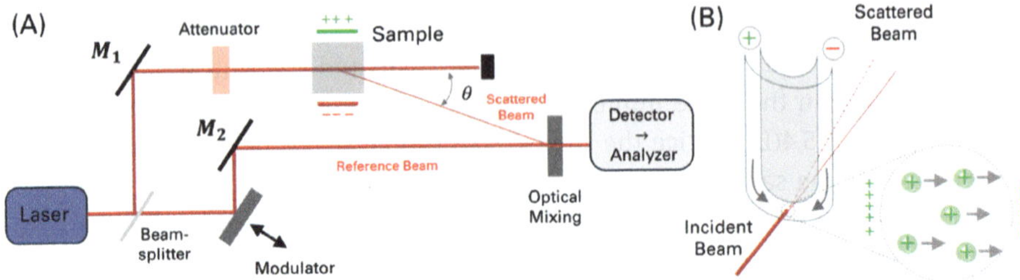

Figure 5.5: (A) Schematic (top view) of a typical electrophoretic light scattering experimental setup. An attenuator is used to modify the power of the laser incident on the sample. The scattered light is typically in the forward direction (13° to 17°). (B) Particles under applied fields in an electrophoretic cell. The beam is focused onto the stationary layer to avoid electroosmosis.

small, difficult-to-resolve Doppler shift into a larger and more easily measurable beat frequency. This significantly increases the sensitivity and thereby the accuracy of ELS measurements. The light scattered from moving particles fluctuates in intensity and frequency over time. The instrument analyzes the frequency spectrum of these fluctuations. The measured Doppler shift is related to the velocity (U) of the particles as follows:

$$\Delta\nu = \frac{qU}{2\pi} = \frac{2n_0 U}{\lambda_0}\sin\left(\frac{\theta}{2}\right) \tag{5.13}$$

where the scattering vector q and the associated optical parameters have been introduced in eq. (5.1). Combining eqs. (5.13) and (3.8), we get

$$\mu = \frac{\Delta\nu \cdot \lambda_0}{2n_0 \cdot \sin\left(\frac{\theta}{2}\right) \cdot |E|} \tag{5.14}$$

This shows that the electrophoretic mobility is directly proportional to the measured Doppler shift. The connection between μ and the zeta potential has been discussed in Section 3.4 for different types of colloids. An estimate of the effective charge (Q_{eff}) of the particle can be obtained from the electrophoretic mobility using the relationship: $Q_{\text{eff}} = 6\pi\eta R_h\mu$.

An innovation in the zeta potential measurement involves phase analysis instead of frequency. In PALS, the same experimental approach as the ELS measurement described is adopted, but the signal processing is different. The phase shift ($\Delta\phi$) between the scattered light and reference beam is measured because it is more sensitive to the changes in mobility compared to the frequency shift. Note, $\frac{d\Delta\phi}{dt} = 2\pi\Delta\nu$. A small change in frequency results in a significant change in phase between the reference and beat frequencies. An electric field is applied (Figure 5.6(A)), and the phase difference is measured and plotted over time (Figure 5.6(B)). The slopes (or the rate of change of phase difference) of these traces are related to the particle electrophoretic mobility.

PALS measurements consist of two different types of reversal of the applied field – a slow field reversal (SFR) and a fast field reversal (FFR). In SFR, the field is reversed slowly (in about every one second) to reduce polarization of the electrodes and allow the fluid flow to stabilize. This approach provides a better resolved distribution of electrophoretic mobilities. However, these values are impacted by electroosmosis. Rapid reversal of the field is performed in FFR mode at the center of the capillary cell. Particles reach a terminal velocity quickly, and this minimizes the influence of electroosmosis. Thus, the FFR method results in a single accurate value of electrophoretic mobility. Information about the distribution is lost because the velocity is sampled over a short period of time. Modern instruments combine both techniques. Figures 5.6(A) and 5.6(B) show the applied field and the corresponding phase measurements, respectively. Both FFR and SFR modes are obtained. The difference between the mean zeta potential from FFR and the mean from SFR is used to determine the electroosmotic flow, which is used to normalize the SFR distribution.

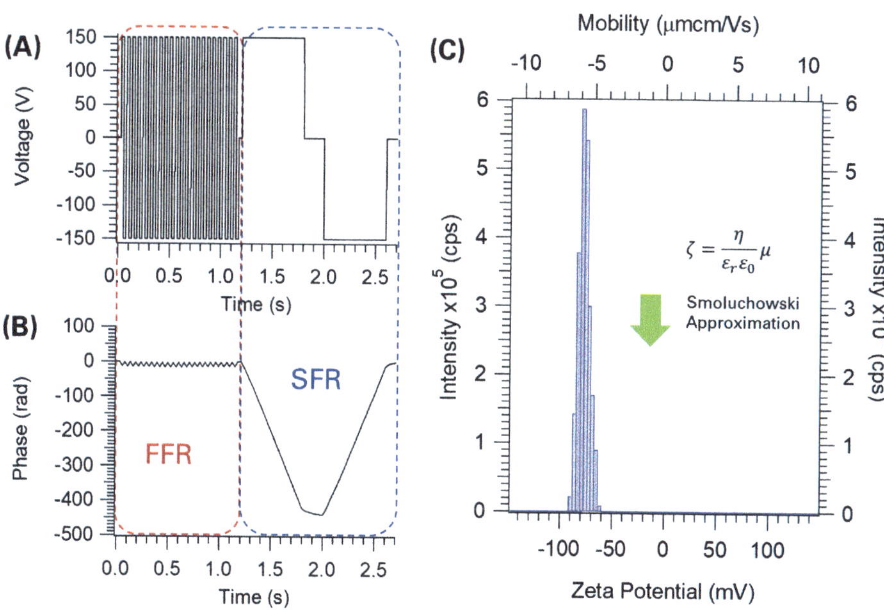

Figure 5.6: The voltage (A) and the phase (B) measurements as a function of time during FFR and SFR field modulations. (C) Example data: electrophoretic mobility and zeta potential distribution.

The PALS method reduces the electroosmosis effect. Additional advantages include averting Joule heating. Since the sensitivity is considerable, the electric field strength can be weak, preventing the electrolysis of the sample and damaging the electrodes. Thus, this technique can be applied to solutions containing high salt concentrations. The high sensitivity also allows for measurement on samples of small volume, as well as accurate measurement on particles of low electrophoretic mobility. Low electrophoretic mobility arises in the case of particles with small charge density, or in a non-aqueous solvent of high viscosity or low dielectric constant. Lastly, FFR mode provides an accurate mean, and the SFR yields a distribution. The reported zeta values show a distribution instead of just a single average value (Figure 5.6(C)).

Zeta potential is not the surface potential. It does not measure surface charge or charge density and only gives an indication of the nature of the surface charge. Numerous assumptions are involved in converting electrophoretic mobility into zeta potential. The particle shape, as well as the interfacial structure, can influence the zeta potential values. Nonetheless, zeta potential is a valuable quantity in colloidal science. It provides a qualitative marker for colloidal stability. In general, a colloidal suspension with zeta potential values greater than ±30 mV is considered stable. Values between ±10 and ±30 mV indicate limited to moderate stability. Particles with a zeta potential close to 0 mV are likely to be unstable and therefore coagulate. However, these values serve as a qualitative guide. Zeta potential reflects colloidal stability from the perspective of Coulombic interaction. Steric and van der Waals interactions also influence colloidal stability (Section 3.5). For instance, if the Hamaker constant is low, the vdW attractive forces are weak. In such a scenario, colloids can remain stable even if they display a low magnitude of zeta potential. Steric repulsions can also ensure colloidal stability even at near zero zeta potential.

Isoelectric point of colloids can be determined by measuring pH-dependent zeta potential. Isoelectric point provides valuable information on the stability of the particle but also offers insights into the nature of the colloidal surface charge. The influence of external factors such as temperature, molecular adsorption, and ionic strength can also be studied based on zeta potential measurements. Titration with ions is necessary to deduce the surface charge density. Further precautions and best practices related to zeta potential measurements are presented in the next section.

5.5 Practical Advice for Experimental Size and Zeta Potential Measurements

These light scattering techniques are powerful as they allow in situ characterization of colloidal substances. However, as we have seen, there are numerous assumptions embedded in determining size and zeta potentials. Various experimental factors also influence data interpretation. Table 5.1 summarizes best practices in size and zeta potential measurement using light scattering.

Table 5.1: Practical recommendations for size and zeta potential measurements.

Best practices – general considerations	
Sample preparation	– Ensure purity of the colloidal suspension – dust particles, aggregates, and air bubbles should be avoided. Use an appropriate filter (or centrifuge) to remove large particles, unless aggregation is the target specimen. Filter pores should be 3× larger than the largest particles of interest. It is important to pre-wet and prime the filter before filtering to remove loose fibers and contaminants, and to ensure a proper seal with the syringe. Filtered dispersant can be used as a blank control. If appropriate, centrifugation can be used as an alternative to filtering. – Prepare an appropriate concentration range – 0.1–1% w/v, or 1–10 mg/mL is a common starting point. It is a good idea to measure the size at different concentrations. Within the diluted range, the measured size should be consistent. – Sample homogenization – for proteins, polymers, or soft fuzzy particles, gentle mixing is ideal and avoids strong agitation, such as sonication. Powders or other sturdy particles can be vortexed and mildly sonicated. To colloidize a solid substance, a solvent that disperses the particles well should be chosen. Surfactants can be used to aid in dispersing solid particles. – Caution must be taken with colored samples – avoid samples that absorb light or fluoresce at the laser wavelength. The use of diluted samples or filters to attenuate or block the light can mitigate the situation.
Experimental setup	– Use appropriate cuvettes – if disposable, make sure the cuvette is solvent compatible. Reusable cuvettes should be cleaned (rinsed, dried) thoroughly. If possible, conditioning the cuvette with the sample is desirable. Ensure no scratches or dust are present on the optical surfaces. – Ensure thermal equilibrium – make sure the sample is fully equilibrated to the set temperature before measurement. Two key inputs, solvent refractive index (n) and viscosity (η), are temperature dependent. Accurate values of n and η must be entered. – Instrument setup – back scattering (~173) is often chosen to avoid multiple scattering; however, this can yield a weak signal. For small particles, side scattering (~90) is ideal. For zeta potential measurement, a detection angle of ~13° is often used.
Best practices – specific recommendations for size measurements	
Data collection	– Monitor the intensity – a good count rate of intensity is 100–500 kcps. Make sure the intensity is stable. – Check the correlogram – the y-intercept should be between 0.85 and 0.95. Weaker scattering or turbid samples will yield a lower y-intercept value. The presence of dust particles can yield a value >1, due to drastic fluctuations. – Perform multiple measurements – it is possible that particles may aggregate or sediment. It is essential to run multiple trials and remix the sample in between to verify accuracy and reproducibility.

Table 5.1 (continued)

Analysis and reporting	– Size distribution – for narrow (PDI < 0.1), monomodal samples, and cumulant analysis is effective. Sizes can be compared with other methods. If PDI is between 0.1 and 0.5, *Z*-average (intensity-weighted mean) values can be used for comparative purposes. For broader distributions of particle size (PDI > 0.5), the use of *Z*-average is not recommended. Multimodal analysis is necessary. It is best to report the intensity-weighted distribution size as the primary results. Volume- and number-weighted distributions involve further assumptions and should be used with caution. – Contextualize DLS data – clearly state the specific experimental parameters, instrumental settings, sample conditions (e.g., pH, ionic strength, temperature, etc.), and the type of analysis when reporting the size and the distribution. For multiple trials, report the mean and error bars.
Best practices – specific recommendations for zeta potential measurements	
Sample condition and quality	– Avoid high salt concentration – could lead to unstable particles, electrode polarization, and Joule heating. – Measure the pH and ionic strength of the sample – they significantly influence particle surface charge and, consequently, zeta potential. – Check electrodes for corrosion before a measurement. Damaged or dirty electrodes interfere with the applied electric field and alter the properties of the sample being analyzed.
Measurements	– Clean thoroughly – use the same solvent as your dispersant for rinsing the cell, and condition the cell with the target sample. – Avoid bubbles – if folded capillary cells are used, load the sample carefully with a syringe to avoid creating air bubbles. Make sure the electrodes are fully submerged. – Make replicate measurements – the reliability of the zeta-potential measurement is at best about ±10% of the measured value. Multiple measurements (>6) and observing the trends in the measured zeta-potential values provide insights into the quality and reproducibility of the data. – Avoid reusing the sample for subsequent zeta potential measurements. Electrophoresis may degrade certain samples.
Reporting electrokinetic properties	– It is best to report not only the zeta potential but also the electrophoretic mobility. The latter is the fundamental quantity directly measured by the ELS technique. For the zeta potential reported, mention the model used. Particle size and surface layer thickness (e.g., polyelectrolytes, conductive surface, etc.) dictate the model used to convert μ to ζ. – Report the pH, ionic strength, particle size and temperature.

5.6 A Survey of Additional Laser Techniques for Colloidal Science

We have reviewed how lasers can be used to study some of the most fundamental (size, molecular weight, and charge) properties of colloids. The light scattering techniques discussed herein fall under the category of linear spectroscopy. Nonlinear spectroscopy, in particular, second-order nonlinear spectroscopy, which is relevant to the study of colloidal interfaces, is discussed in Chapter 6. However, the use of lasers and the light scattering approach in colloidal science is much broader. Although a detailed presentation of all these techniques is beyond the scope of this book, in this section, additional applications of lasers in colloidal science are briefly surveyed.

5.6.1 Ancillary Laser Instruments for Particle Size and Shape

Laser diffraction is a technique (based on the principles of SLS) that allows particle size distribution measurement. Compared to DLS, it can measure a much wider range of particle sizes – spanning less than a hundred nanometers to a few millimeters in size. A series of detectors, typically arranged in a semicircle, measures the diffraction pattern created by the suspended particles. Fraunhofer, applicable for particles larger than the wavelength of the light, and Mie theories are employed. As we have seen, the angular pattern of the diffracted or scattered intensity is sensitive to the particle size. The size distribution is thus obtained from the sum of the diffraction patterns. Typical light sources are diode lasers in the visible (red or blue) wavelength range.

Multiangle laser light scattering, MALLS, can yield qualitative information about the particle shape. We have seen that SLS techniques yield the radius of gyration of the macromolecule. The shape factor is defined as the ratio of R_g to R_h. The latter quantity, the hydrodynamic radius, is determined based on DLS. Generally, cylinders or rod-like structures can have $\frac{R_g}{R_h} \gtrsim 2.0$ and random coils, a value of 1.0–1.5. For compact solid, the $\frac{R_g}{R_h} = 0.775$. High values indicate a nonspherical or extended shape. An extension of DLS, known as depolarized DLS, utilizes polarized and depolarized laser lights. Depolarized DLS allows the simultaneous determination of the translational and rotational diffusion coefficients by separating the scattered light into vertically and horizontally polarized components. These two coefficients are used to calculate the aspect ratio of elongated particles, such as rods or ellipsoids.

The elliptically polarized light scattering technique uses the polarization state of scattered light to characterize particle shape, size, and aggregation. It involves an instrument that can both generate a controlled elliptically polarized incident beam and analyze the resulting scattered light. By analyzing the asymmetrical nature of scattered light, particle aspect ratios are measured. Differentiation between spheres and nonspherical shapes like ellipsoids or aggregates is thus possible. Elliptically polarized light (see Chapter 4) offers detailed information about the particle's morphology because it is sensitive to asymmetries in the particle's structure.

5.6.2 Light Scattering for Particle Detection, Tracking, and Separation

The light scattering techniques discussed are useful in detecting and tracking different types of particles. Nanoparticle tracking analysis employs lasers to illuminate coherent light on colloids. The scattered light is captured by a sensitive camera to record the Brownian motion of moving particles in real time against a dark background. Lasers of different wavelengths and intensities can be used to analyze a broad range of particle sizes. The technique can also incorporate excitation of fluorophores to detect fluorescently labeled particles. The particle tracking algorithm tracks the movement of individual particles in a video sequence. Nanoparticle tracking analysis allows the determination of size, size distribution, and concentration of particles within the colloidal range.

Light scattering mechanisms are also combined with separation or flow fractionation techniques discussed in Chapter 4. In a laser-based field-flow fractionation method, MALLS detectors are used to measure the size distribution and concentration of separated components of a colloidal sample. A flow-channel system separates samples into fractions based on their diffusion rates, with larger particles moving slower. MALS then analyzes these fractions to determine the particle size and mass, providing quantitative information on the flow field fraction for different particle sizes. This technique is valuable for a wide range of macromolecules, including proteins, viruses, nanoparticles, and other colloids.

Particle tracking is not only possible in laboratory settings. Light detection and ranging (LiDaR) is a technique that uses laser pulses to sense and map aerosol particles and clouds in the atmosphere [81]. It is a remote sensing method that measures the time it takes for backscattered light to return. It works on the principles of DLS and laser diffraction. The intensity of the backscattered light, along with the time delay, indicates the density and type of aerosols present. Doppler shifts can also reveal wind speed and direction. Beyond size, LiDaR can also provide information on the complex refractive index and shape of the atmospheric aerosols. Aerosol vertical profiling, estimating their microphysical properties and interactions with clouds, are all important factors for climate modeling, weather forecasting, and monitoring air quality and pollution.

5.6.3 Laser Techniques to Study Colloidal Composition

Various other laser-based techniques used for molecular and elemental analysis have also been applied to study colloids. In laser-induced fluorescence (LIF), a laser beam excites fluorescent molecules or particles within the colloidal system. The use of a laser ensures highly controlled and precise excitation. A higher spatial resolution and sensitivity are achieved. The emitted photons are detected and can be analyzed to determine the concentration, distribution, and dynamics of colloidal particles. Fluorescence serves as a marker to detect and track the particles; thus, LIF allows real-time visualization of how colloids flow through different media and their aggregation or deposition over time.

Unlike LIF, laser-induced breakdown or plasma spectroscopy is often a destructive, nonsampling analytical technique. It involves a process known as ablation, in which a high-energy pulsed laser beam is focused on a sample's surface, causing a small amount of substance to vaporize and form a high-temperature plasma. This plasma is comprised of ionic species and excited atoms that emit light at specific wavelengths unique to the elements present in the substance. Accordingly, laser-induced breakdown spectroscopy provides information on the elemental composition of materials. A closely related method is laser-induced breakdown detection. The pulse energy of the laser is set below the breakdown threshold of the pure solvent, but above the threshold of the solid particles. The plasma produced is detected in two ways – optical versus acoustic detection. In the latter case, piezoelectric sensors detect the shock wave caused by the plasma expansion. The amplitude of the acoustic signal is related to the particle size. Imaging the spatial distribution of plasma events using a CCD camera is used to infer particle size. Larger particles generate plasma over a larger area. Laser-induced breakdown detection allows detection of very low and wide concentrations of colloids (1 ng/L to mg/L). Both mean particle size and concentrations can be determined. It is more sensitive to small particles than conventional light scattering methods.

5.6.4 Surface-Enhanced Raman Spectroscopy and Colloids

Chapter 4 introduced Raman spectroscopy. Vibrational Raman spectroscopy has an array of applications in many areas, including colloidal science, as it allows the identification of the organic composition of materials. A special form of Raman spectroscopy that is valuable in colloidal chemistry is known as surface-enhanced Raman spectroscopy, or SERS. It involves significant amplification of Raman signals that are adsorbed onto or near a metallic substrate. Generally, metallic nanoparticles such as gold and silver are used. The primary mechanism by which enhanced Raman scattering occurs is called localized surface plasmon resonances (LSPR). When light interacts with the conduction electrons of small metallic particles, LSPR is created, which is essentially a "hotspot" of immense concentration of electromagnetic fields at the nanoparticle surface. The Raman signal from molecules present at this hotspot is dramatically amplified due to the localized field. Signal amplification in the order of 10^{10}–10^{14} can occur. Thus, SERS is highly sensitive and can offer single-molecule detection capability.

Another mechanism involves the formation of a temporary charge-transfer state that alters the molecule's polarizability, leading to enhanced Raman scattering. This mechanism of chemical enhancement is less powerful than the LSPR enhancement.

Graphene and its derivatives can act as SERS substrates via chemical enhancement. Since graphene also quenches fluorescence, the signal quality of the SERS is enhanced. Additionally, polymers can be decorated with metallic nanoparticles. Such engineered substrates can still provide the primary plasmonic enhancement.

Given that colloidal nanoparticles are integral in the process of SERS, this technique is uniquely suited for studying colloids, and particularly their surfaces. SERS provides detailed information about nanoparticle-bound species. This includes investigation of how different molecules bind to the colloid, their orientation at the interface, and surface interactions. Kinetics of aggregation of colloidal particles in real time have also been studied using SERS. SERS spectra can yield information about the shape, size, and aggregation state of the nanoparticles. Nanoparticles with sharp edges, such as nanostars or nanorods, provide intrinsic hot spots and stronger SERS signals than isolated spheres. While highly sensitive, the application of SERS to colloids exhibits several limitations related to reproducibility, stability, and quantification. These challenges stem from the dynamic and complex nature of colloidal nanoparticle systems, which create variability in the SERS-enhancing hot spots. Uncontrolled aggregation leads to SERS intensity fluctuations. Variation in sample preparation, changes in environmental factors (e.g., pH, temperature, ionic strength) can significantly modify the efficacy of the enhancement.

5.6.5 Engineering and Synthesis of Colloids Using Lasers

Lasers are not only used to characterize colloids. They also offer an approach to synthesize and engineer precisely controlled colloidal materials [82–86]. High-purity and ligand-free colloids can be prepared from a wide array of materials. Collectively, this approach involving scalable laser-based methodologies is called laser synthesis and processing of colloids [85, 86]. Some of these techniques are known as pulsed laser ablation in liquid (PLAL), laser fragmentation in liquids (LFL), pulsed laser melting in liquids (PLML), laser-induced reduction in liquids (LRL), and laser defect engineering in liquids. The basic premise is that various lasers are used for ablation, fragmentation, and melting within liquid environments.

PLAL is a top-down approach, in which a pulsed laser is focused onto the target bulk material submerged in a liquid. A high-energy plasma plume is created, which rapidly expands, cools, and condenses to form nanoparticles. Particle size is influenced by the pulse duration and fluence (energy per area) of the laser. Femtosecond lasers at low fluence can produce small, monodispersed particles. Nanosecond pulses lead to larger particles with broader size distribution. The LFL approach is to break down larger particles into smaller ones by irradiating a colloidal dispersion. This technique can achieve particles as small as sub-5 nm in diameter. The melting of small particles and aggregating them to larger spherical particles is the main mechanism for PLML. Rapid cooling by the surrounding liquid quenches the particles into single-crystalline spheres. Particle size can be tuned by controlling the laser fluence and irradiation time. Laser irradiation of a solution containing metallic ions is performed in LRL. Reduction of metal ions leads to the formation of nanoparticles. The basis of laser Doppler electrophoresis is to introduce and manipulate structural defects or dopants using controlled laser irradiation.

When it comes to manipulating colloids, the method of optical tweezers is a versatile tool [87, 88]. It involves a tightly focused laser beam to create radiation pressure and gradient forces to trap, hold, and move microscopic particles. Trapping the particle has a lot of applications in colloidal engineering. For example, optically trapped particles can be precisely positioned to build a microstructure in three-dimensions. They can be guided to self-assemble into specific configurations of interest. In addition to microstructure engineering and assembly, optical tweezers are also useful to determine fundamental properties of colloids, such as colloidal mechanics and interactions. Optical tweezers can be used to measure the electrostatic and van der Waals forces between individual particles. The effect of the fluid environment and external stimuli, such as temperature and pH, on these forces provides insight into the physicochemical properties of various colloids. Microrheology is another area of study where colloids in combination with optical tweezers provide fundamental information about the viscoelastic properties of complex fluids. By manipulating microparticles within a fluid medium and measuring their mechanical response, rheological properties such as viscosity and elasticity of the material are ascertained.

This brief survey highlights the prevalent use of lasers in colloidal science and technology. It is not just for characterization; lasers are also used for the synthesis of colloids and the development of specialized nanostructures using colloids. The laser-based light scattering methods discussed thus far predominantly apply to characterizing bulk physicochemical properties of colloids. In the next chapters, we explore how lasers are used to study colloidal interfaces directly.

End-of-Chapter Problems

1. Concepts, definitions, and critical thinking:
   (A) Explain the difference between SLS and DLS.
   (B) What type of molecular weight average is measured by light scattering (as compared to osmometry) in the case of a polydisperse suspension?
   (C) Other than measuring colloidal properties, how are lasers used in the realm of colloidal science?

2. In a DLS experiment, a laser is used to measure the Brownian motion of colloidal particles and determine their size. SLS measures the time-averaged intensity of scattered light at various angles.

(A) How does the size of the particles relate to their diffusion speed?
(B) How is the hydrodynamic radius calculated from the measured data?
(C) What specific properties of colloids or macromolecules can be determined using SLS?
(D) What colloidal property can be measured using ELS? When characterizing colloidal stability using light scattering methods, how does the ionic strength of the solution affect the electrostatic repulsion between charged colloidal particles?

3. A DLS experiment is performed on a suspension of monodisperse polystyrene nanoparticles in water at 25 °C. The instrument uses a He-Ne laser and a scattering angle of 173°. The software performs a cumulant analysis and provides a decay constant $\Gamma = 6,500\ s^{-1}$. Calculate:
   (A) the scattering vector
   (B) translational diffusion coefficient
   (C) the hydrodynamic diameter (or Z_{avg})
   (D) the PDI, given that the DLS software uses cumulant analysis and reports a k_2 value of $5 \times 10^6\ s^{-2}$.

4. How do the intensity, volume, and number distributions of a DLS particle size analysis differ, and under what circumstances is each distribution most appropriate for reporting particle size? What assumptions are made regarding the optical properties, shape, and composition of particles in converting the primary DLS data (an intensity-weighted distribution) into volume-weighted or number-weighted distributions, and why are these derived distributions typically considered relative rather than absolute?

5. A macromolecule comprised of cellulose nitrate is dissolved in acetone. Light-scattering data at various scattering angles are collected at different concentrations of cellulose nitrate. A Zimm plot is created, and the following information [57] is obtained:
   The slope of the $\theta = 0$ line is 5.70×10^{-7} mol/g
   Slope of the $c = 0$ line is 6.78×10^{-6} mol/g
   Intercept = 7.87×10^{-7} mol/g
   Determine the molecular mass, second virial coefficient, and the radius of gyration of cellulose nitrate.

6. A PALS measurement is performed on a colloidal suspension of polystyrene nanoparticles in an aqueous medium at 25 °C to determine their zeta potential. The instrument measures the electrophoretic mobility of the particles to be $3.5 \times 10^{-8}\ m^2/Vs$.
   (A) Calculate the zeta potential using the Smoluchowski limit equation.
   (B) Calculate the zeta potential using the Hückel limit equation.

(C) Compare the results and determine which limit is more appropriate for this system if the experiment was conducted in a high-salt (e.g., 10 mM NaCl) aqueous solution.

7. On the best practices of size and zeta potential measurements using DLS
   (A) Why is the presence of dust or large aggregates the most common problem in DLS, and how can it be avoided during sample preparation?
   (B) What is the best practice for filtering samples and solvents?
   (C) Why is temperature control important?
   (D) What is the optimal sample concentration for DLS-based measurements?
   (E) What is a correlogram, and why is it important to inspect the raw data instead of just the final size result?
   (F) What type of medium/solvent should be used for sample preparation, especially regarding ionic strength, for zeta potential measurement?
   (G) How important is sample preparation regarding contamination and air bubbles, and how can these be avoided?
   (H) If the standard reference material is giving an unexpected value (e.g., −50 mV instead of −70 mV), what should be checked?

8. Laser synthesis and processing:
   (A) Describe the process of pulsed laser synthesis and processing of colloids in liquids and its primary advantages over traditional chemical methods.
   (B) Highlight the differences in PLAL, PLML, LFL, LRL, and laser defect engineering in liquid colloidal synthesis techniques.
   (C) In laser fragmentation of colloidal gold nanoparticles, what critical parameter determines if the fragmentation is a single-pulse, one-step event that yields particles less than 10 nm?
   (D) What are optical tweezers (laser tweezers), and what fundamental principle allows them to trap and manipulate colloidal particles?

6 Fundamentals of Nonlinear Spectroscopy and Colloids

In Chapter 4, absorption and emission spectroscopy have been discussed along with different types of light-scattering theories. In all these processes, the light-matter interaction involves one primary incident electromagnetic field. In the particle picture of light, we can think of this as one photon in and one photon out! The intensity of the radiation is weak, and it only induces a linear response in the material. That is, the material's response is directly proportional to the strength of the applied EM field. This type of process falls under the category of linear spectroscopy. The light scattered, absorbed, or emitted is linearly related to the intensity of the incident light. UV-Vis, infrared (IR), fluorescence, and phosphorescence are all examples of linear spectroscopy. Herein, the basic aspects of nonlinear spectroscopy (NLS) are presented.

6.1 Linear Versus Nonlinear Spectroscopy

When an EM field ($\vec{E}$) interacts with matter, it results in an induced polarization ($\vec{P}$) in the material – a distorted charge distribution in the atoms or molecules that make up the material. The quantity of $\vec{P}$ is thus the electric dipole moment per unit volume induced as the electric field displaces the electrons within the substance. Mathematically, $\vec{P}$ can be expressed as a Taylor series expansion in terms of the applied electric field as follows:

$$\vec{P} = \vec{P}_{\text{dc}} + \varepsilon_0 \left(\chi^{(1)} \cdot \vec{E} + \chi^{(2)} : \vec{E}\vec{E} + \chi^{(3)} \vdots \vec{E}\vec{E}\vec{E} + \cdots\right) \tag{6.1}$$

It is represented as a vectorial sum of terms, each representing a different order of the electric field. The first term is the frequency-independent component. It is called the DC (direct current) term and represents the residual or initial polarization that remains after the external field is removed. The second term, which can be separated as $\vec{P}^{(1)} = \varepsilon_0 \chi^{(1)} \cdot \vec{E}$, is first-order polarization. It describes induced polarization as a linear function of the applied field and is the basis of linear spectroscopy. This can be visualized in Figure 6.1(A), which shows the variation in $\vec{P}$ with $\vec{E}$. The remaining terms collectively represent the nonlinear polarization terms and describe nonlinear interactions of light with matter. For example, $\vec{P}^{(2)} = \varepsilon_0 \chi^{(2)} : \vec{E}\vec{E}$ describes the second-order NLS, and $\vec{P}^{(3)} = \chi^{(3)} : \vec{E}\vec{E}\vec{E}$, third-order NLS. The symbol $\chi^{(1)}$ is known as the first-order susceptibility, and the rest ($\chi^{(2)}$, $\chi^{(3)}$, etc.) are the higher-order susceptibilities. These macroscopic susceptibility coefficients, $\chi^{(i)}$, are related to the molecular polarizabilities of the material. They represent the intrinsic optical property of a material substance (or a molecule) describing the tendency of its dielectric to polarize in response to an electromagnetic field.

 | https://doi.org/10.1515/9783111018386-006

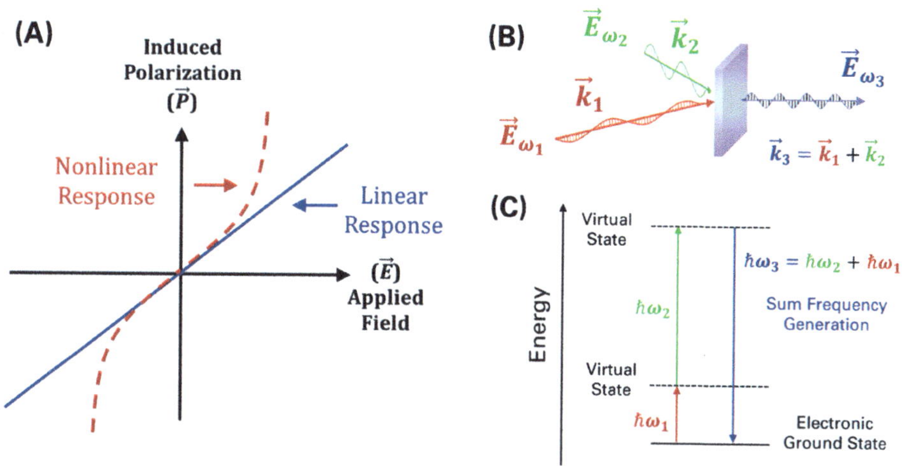

Figure 6.1: (A) Cartoon depicting nonlinear versus linear response of materials under applied electric field. (B) Coherent nonlinear interaction of EM fields (three-wave mixing) with a material. (C) Energy diagram depicting the SFG process. The length of the SFG arrow is the sum of the two arrows representing the incident fields.

It is important to note that the expansion (eq. (6.1)) holds for weak to moderately strong electric fields, and the relationship $\chi^{(1)} \cdot \vec{E} \gg \chi^{(2)} \colon \vec{E}\vec{E} \gg \chi^{(3)} \vdots \vec{E}\vec{E}\vec{E}, \ldots$ must hold for the series to converge. Typical magnitudes [89] of $\chi^{(2)}$ and $\chi^{(3)}$ for condensed matter are on the order of $10^{-12}\,\frac{\mathrm{m}}{\mathrm{V}}$ and $10^{-19}\,\frac{\mathrm{m}^2}{\mathrm{V}^2}$, respectively. Compared to $\chi^{(1)}$ values in the order of $1-10^3$, the higher order susceptibility values are very small. Ordinary household lights have power in the range of a few tens to a few hundreds of watts. Thus, nonlinear optical effects are not readily observed under conventional light sources. Nonlinear effects become possible when the light intensity is strong, as is often achieved by a laser as the light source. As discussed in Chapter 4, a Tsunami laser can produce a peak power of 10^4 watts and when focused, a peak intensity in the order of $10^{14}\,\frac{\mathrm{watts}}{\mathrm{m}^2}$.

Let us take a simplistic approach and consider the resulting induced polarization when a linearly polarized EM wave interacts with the material system [89, 90]. The field is oscillating at a frequency ω and propagating in the z direction. We can describe the EM field as $E = E_{\mathrm{dc}} + E_0 \cos(\omega t - kz)$, where E_{dc} is the DC component of the field and E_0, the electric field amplitude oscillating at ω. Plugging E into eq. (6.1), we have

$$\vec{P} = \vec{P}_{\mathrm{dc}} + \varepsilon_0 (\chi^{(1)} [E_{\mathrm{dc}} + E_0 \cos(\omega t - kz)] + \chi^{(2)} [E_{\mathrm{dc}} + E_0 \cos(\omega t - kz)]^2 + \chi^{(3)} [E_{\mathrm{dc}} + E_0 \cos(\omega t - kz)]^3 + \cdots)$$

Upon expansion and applying appropriate trigonometric identities, we get

$$\vec{P} \propto \varepsilon_0 (\chi^{(1)} E_{\mathrm{dc}} + + \chi^{(1)} E_0 \cos(\omega t - kz) + \chi^{(2)} E_{\mathrm{dc}}^2 + 2\chi^{(2)} E_{\mathrm{dc}} E_0 \cos(\omega t - kz) + \frac{1}{2}\chi^{(2)} [E_0^2 + E_0^2 \cos(2\omega t - 2kz)] + \cdots)$$

Additional terms generated in this expansion are provided in Table 6.1. We can see that the linear response (terms associated with $\chi^{(1)}$) leads to an induced polarization having components oscillating at ω, which is the same frequency as the incident light wave. However, it is evident that in the presence of strong electric fields, the nonlinear response (e.g., $\chi^{(2)}$) of the medium leads to the generation of harmonics (e.g., 2ω) in $\vec{P}$. In Table 6.1, higher harmonics, such as induced polarization oscillating at 3ω, are also shown.

Table 6.1: Select first-, second-, and third-order processes.

Terms	Frequency of the response	Electro-optical phenomenon
$\chi^{(1)} E_{\mathrm{dc}}$	DC	DC polarizability
$\chi^{(1)} E_0 \cos(\omega t)$	ω	Optical polarizability (refractive index)
$\chi^{(2)} E_{\mathrm{dc}}^2$	DC	DC hyper-polarizability
$\chi^{(2)} E_{\mathrm{dc}} E_0 \cos(\omega t)$	ω	Linear electrooptical effect (Pockel's effect)
$\chi^{(2)} E_0^2$	DC	Optical rectification
$\chi^{(2)} E_0^2 \cos(2\omega t)$	2ω	Second harmonic generation (SHG)
$\chi^{(3)} E_{\mathrm{dc}}^2 E_\omega \cos(\omega t)$	ω	DC Kerr effect
$\chi^{(3)} E_{\mathrm{dc}} E_0^2 \cos(2\omega t)$	2ω	DC-induced SHG
$\chi^{(3)} E_0^3 \cos(3\omega t)$	3ω	Third harmonic generation (THG)
$\chi^{(3)} E_0^3 \cos(\omega t)$	ω	Optical frequency Kerr effect
⋮	⋮	⋮

The induced polarization acts as the source term in the Maxwell equation [89, 91]. Equation (6.2) is applicable under the slow varying amplitude approximation and further assumes that the material is nonmagnetic and the region under consideration contains no free charges, i.e., $\rho = 0$:

$$\nabla^2 \vec{E} - \mu_0 \varepsilon_0 \frac{\partial^2 \vec{E}}{\partial t^2} = \mu_0 \frac{\partial^2 \vec{P}}{\partial t^2} \tag{6.2}$$

A key concept revealed in this equation is that as the induced dipole moment oscillates in time, it radiates energy in the form of light. The corresponding electric field, $\vec{E}$, is shown on the left side of the equation. This newly generated field is directly proportional to the induced polarization. That is, if $\vec{P}$ oscillates at 2ω, the generated radiating light will also oscillate at 2ω. This process is called second harmonic generation or SHG. Similarly, light generated at 3ω due to nonlinear interactions is called third harmonic generation. Certainly, higher harmonics are possible with increasing intensity of the fundamental laser beam.

In the example above, the incident fields interacting with the nonlinear medium have the same frequency. Now imagine, EM fields of two different frequencies – one oscillating at ω_1 and the other at ω_2 – are impinged upon a material as depicted in Figure 6.1(B). Nonlinear interaction of two EM waves results in an output field oscillating at ω_3. Depending on the phase matching conditions (see Section 6.2), the resulting frequency can either be the sum of ($\omega_3 = \omega_1 + \omega_2$) or the difference ($\omega_3 = |\omega_1 - \omega_2|$) between the two original frequencies. The former is known as sum frequency generation (SFG) and the latter, difference frequency generation (DFG). In either scenario, energy is conserved. This means $\hbar(\omega_1 + \omega_2) = \hbar\omega_3$. Equation (6.3) allows calculation of the SFG wavelength (λ_{SFG}), where λ_1 and λ_2 are the wavelengths of the input laser beams:

$$\lambda_{SFG} = \frac{\lambda_1 \lambda_2}{\lambda_1 + \lambda_2} \tag{6.3}$$

As shown in Figure 6.1(C), a net energy transfer occurs from the input beams to the newly generated higher-frequency beam in the process of SFG. The nonlinear material converts the input photons into a newly generated higher-energy photon. In DFG, the nonlinear interaction of one photon leads to the generation of two photons. The output photons are lower in energy compared to the incident photon. In general, the simultaneous occurrence of SFG and DFG is rare, and only one process is optimized at a time.

Collectively, SFG, DFG, and SHG are known as parametric processes, in which the quantum states of the materials remain unchanged upon interaction with light. In other words, there is no direct energy, momentum, or angular momentum transferred to or from the system. For example, in the realm of linear optics, elastic scattering (e.g., Rayleigh and Mie scattering), refraction, and diffraction are parametric, lossless dielectric processes. In contrast, optical absorption, Raman scattering, and emissions are nonparametric processes. Unlike SFG light scattering, two-photon absorption is a nonparametric process in the realm of second-order nonlinear optics. For parametric processes, the elevated energy level is represented as a temporary virtual state in the energy diagram (Figure 6.1(C)). The virtual state is transient (i.e., short-lived), unstable, and an intermediate state. Based on the Heisenberg uncertainty principle, the lifetime (Δt) of the parametric process involving virtual states is in the order of $\Delta t \sim \frac{h}{2\pi\Delta E}$, where ΔE is the difference in energy states and h is Planck's constant. If the SFG process is in the visible range, the lifetime is roughly 10^{-16}s, far shorter than electronic transition processes within an atom or molecule.

6.2 Nuts and Bolts of Second-Order NLS

The nonlinear processes are essential in producing tunable sources of coherent light in the infrared, visible, and terahertz regions. More importantly, these phenomena rely on a unique selection rule: they occur in a region within the nonlinear medium that exhibits broken inversion symmetry. This criterion allows these processes to be selective for interfacial studies, which is a major focus of this book. SHG is a special case of SFG, in which the incident photons have degenerate frequencies. SFG is a general phenomenon and includes the special case of SHG as well. The frequency doubling process of SHG was first demonstrated by Franken et al. [92] in 1961 using a monochromatic light source from a ruby laser. The applicability of SHG and SFG for interfacial studies became thoroughly recognized later in the 1980s, with the first SHG experiment carried out on an adsorbed layer in 1969. We begin with the basic principles of the second-order NLS phenomena, as they were developed for planar interfaces. The significance of interfacial selectivity and the investigation of colloids using SHG and SFG will follow.

6.2.1 The Second-Order Susceptibility and Interfacial Selectivity

Our focus is the second-order nonlinear induced polarization, $\vec{P}^{(2)}$. For extensive theoretical descriptions of NLS and its extension to surface SFG and SHG, readers are encouraged to refer to the following books [89, 91, 93]. Herein, the key concepts of these processes are outlined. The general form of $\vec{P}^{(2)}$ can be expressed as follows:

$$P_i^{(2)}(\omega_3 = \omega_1 + \omega_2) \propto \sum_{j,k} \chi_{ijk}^{(2)}(\omega_3;\omega_1,\omega_2){:}E_j(\omega_1)E_k(\omega_2) \tag{6.4}$$

If the frequencies $\omega_1 = \omega_2$, we have SHG. Equation (6.2) dictates that the SFG or SHG field, $E_i^{(2)}(\omega_1 + \omega_2)$ or $E_i^{(2)}(2\omega)$, generated from the interaction, will be proportional to $P_i^{(2)}$. The indices, i, j, k, represent the Cartesian coordinates $(x,\ y,\ z)$ and the frequencies $(\omega_3;\omega_1,\omega_2)$ are in reference to the output and the input EM fields. Recall, $P_i^{(2)}(\omega_3)$ is proportional to $E_i(\omega_3)$. The first index i describes the direction of the second-order-induced polarization vector in Cartesian coordinates. The two input fields get the j and k designations. Given the vectorial nature of the fields and the induced polarization, the susceptibility term connecting the two is therefore a tensor. A tensor describes multilinear relations between scalar quantities, geometric vectors, matrices, or other tensors. Specifically in this case, $\chi_{ijk}^{(2)}$ is a third-rank tensor and has a total of 27 elements. For comparison, a scalar quantity has one element, a vector has three elements, and a 3x3 matrix, nine elements. The total induced $P_i^{(2)}$ and thereby the overall sum frequency signal is thus the summation of the contributions from all these elements as shown in eq. (6.4).

However, various criteria based on symmetry render many of the susceptibility elements equal to zero, meaning not all $\chi_{ijk}^{(2)}$ terms contribute to the signal. In fact, a concept from group theory, known as Neumann's principle, dictates that the symmetry elements (e.g., reflections, rotations, and inversions) that leave the crystal or molecular structure unchanged must also leave its physical property (e.g., second-order susceptibility) unchanged. Accordingly, a molecule exhibiting the lowest symmetry point group, C_1, would exhibit contribution from all 27 elements to the overall SFG intensity. The C_1 point group only possesses the identity symmetry element (i.e., rotation by 360°). For collinear SHG, the two incoming fields have the same frequency and are aligned in the same direction. Thus, $\chi_{ijk}^{(2)}$, in the case of SHG, is symmetric in its last two indices (j and k) and the number of independent elements is reduced. For C_1 point group, the contribution reduces to 18 $\chi_{ijk}^{(2)}$ terms. For crystals or molecules belonging to the higher symmetry point groups, the independent susceptibility elements are further reduced. For instance, there are only seven nonvanishing $\chi_{ijk}^{(2)}$ elements for molecules belonging to the C_{2v} point group, and they are $\chi_{zzz}^{(2)}$, $\chi_{zxx}^{(2)}$, $\chi_{zyy}^{(2)}$, $\chi_{yzy}^{(2)}$, $\chi_{yyz}^{(2)}$, $\chi_{xzx}^{(2)}$, and $\chi_{xxz}^{(2)}$.

A major takeaway is that these $\chi_{ijk}^{(2)}$ terms reflect the material's symmetry and the polarization dependence of interaction. Therefore, the measured $\chi_{ijk}^{(2)}$ values from the SHG or SFG intensities can provide information about the crystal structure and, in the case of adsorbed molecules at an interface, their average orientation with respect to the surface plane. Neumann's principle has a big impact on second-order nonlinear processes when inversion symmetry is considered. The inversion operation ($\hat{\imath}$) changes the sign of all spatial coordinates as follows: $\hat{\imath}(x, y, z) \rightarrow (-x, -y, -z)$). Both the induced polarization and the EM fields are vectors, and upon inversion, they change signs from $\vec{P}^{(2)}$ to $-\vec{P}^{(2)}$ and $\vec{E}=$ to $-\vec{E}$. However, according to Neumann's principle, $\chi_{ijk}^{(2)}$ elements for a centrosymmetric material do not change sign. The term "centrosymmetric" means that the object remains unchanged when inverted through its center. Thus, applying the inversion operator on eq. (6.4), we get

$$-P_i^{(2)} \propto \sum_{j,k} \chi_{ijk}^{(2)}:[-E_j(\omega_1)][-E_k(\omega_2)]$$

$$-P_i^{(2)} \propto \sum_{j,k} \chi_{ijk}^{(2)}:E_j(\omega_1)E_j(\omega_2)$$

This result is a contradiction unless all $\chi_{ijk}^{(2)}$ elements are zero. Thus, we conclude that centrosymmetric materials (or an isotropic medium in which properties of materials are the same in all directions, regardless of the orientation in which they are measured) do not generate a second-order nonlinear response. Thus, within the dipole approximation, coherent SFG is not allowed in an isotropic medium. In fact, this selection rule is applicable to any even-order nonlinear processes.

In contrast, anisotropic materials, where centrosymmetry is broken, do not have such a restriction. Most chemical interfaces are inherently anisotropic. Molecules at an interface experience different forces that vary with direction. SFG is allowed at chemical interfaces surrounded by a bulk isotropic medium. As a result, SFG serves as a powerful tool to study chemical interfaces selectively. In other words, SFG signals do not include contributions from molecules in bulk surroundings but only from the interfacial bound species. It is important to note, however, that if the material is anisotropic, the bulk material can yield a signal at the sum frequency, and surface specificity is convoluted in this case. Mechanisms such as dipole fluctuations, electric field induced, and quadrupolar contributions from the material can result in a sum frequency response. Consequently, SFG is not always surface specific. However, manipulating experimental parameters allows distinguishing interfacial-originated SFG signals from those of a different mechanism (see Sections 6.3 and 6.4). Nevertheless, in the backdrop of an isotropic material, as is the case for amorphous particles suspended in an isotropic dispersant, SFG serves as a unique tool that cuts through the bulk and retrieves information from the molecularly thin interfacial layer.

6.2.2 SFG Intensity and Phase Matching Condition

Intensity is proportional to the square of the radiating electric field (eq. (4.2)). Accordingly, the overall SFG intensity is expressed as

$$I_{SFG} \propto |E_{SFG}|^2 \propto \left|P^{(2)}_{SFG}\right|^2 \tag{6.5}$$

I_{SFG} is the experimentally measured quantity, and it depends not only on the molecular parameters but also on various experimental factors. For instance, if a planar surface is probed in the reflection geometry, the I_{SFG} efficiency will depend on the angles of incidence of the fundamental beams. When scattered SFG from a colloidal suspension is detected, the number density of the particles, the pathlength of the light propagation through the sample, and the angle at which the signal is detected all play a role. The strength of the SFG signal is also dependent on the polarizations of the incident and the detected SFG light.

In a bulk anisotropic medium, an important criterion that dictates the efficiency of SFG is the phase-matching condition. It is based on the law of conservation of momentum. The photons involved must conserve their momentum, as shown in Figure 6.1(B). As a result, the phase matching condition is related to the wavevectors as follows: $\Delta\vec{k} = \vec{k}_3 - \vec{k}_1 - \vec{k}_2 = 0$. When $\Delta\vec{k} \neq 0$, there is a phase mismatch. Phase matching or constructive interference occurs when the incoming and the output waves maintain the same relative phase as they propagate through the medium. Recall, the wavevector is related to the refractive index $\left(k = \frac{2\pi n}{\lambda}\right)$, and the refractive index varies with wavelength. Given that the wavelength of the SFG field is different from that of the fundamental fields, it is not straightforward to maintain the phase-matching condition over a long distance of light propagation. Equation (6.6) expresses SFG intensity as a function of the product of the propagation distance (z), and the phase parameter ($\Delta\vec{k}$):

$$I_{\text{SFG}}(\omega_3) = \frac{8\omega_3^2}{n(\omega_1)n(\omega_2)n(\omega_3)\varepsilon_0 c^2}\left|\chi_{\text{eff}}^{(2)}\right|^2 I(\omega_1)I(\omega_2)\text{sinc}^2\left(\frac{\Delta k \cdot z}{2}\right) \tag{6.6}$$

The intensity oscillates with the maximum possible efficiency obtained at $\Delta k = 0$, and drastically diminished intensity as the field propagates through the medium (Figure 6.2 (A)). There is a special distance, called coherence length (L_c), over which the fundamental and the SFG fields remain in phase. It is defined as $L_c = \frac{\pi}{|\Delta k|}$. Beyond this distance, the phase difference becomes large, and destructive interference causes SFG intensity to decrease. The concept of interference and phase relationship will become important when we discuss SFG originating from colloidal interfaces in Section 6.3.

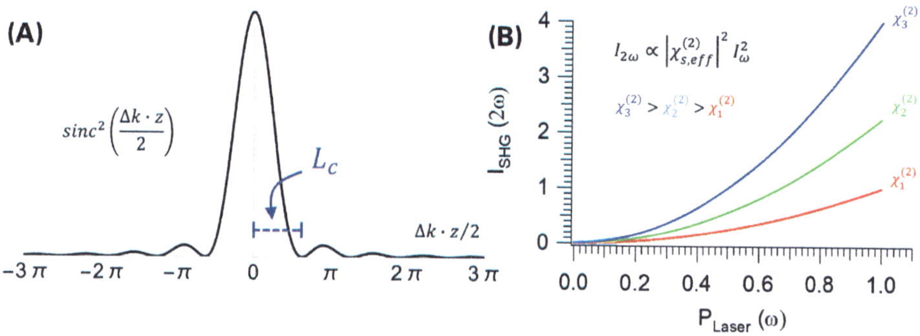

Figure 6.2: (A) SFG intensity profile as a function of the phase mismatch. (B) SHG intensity as a function of the fundamental laser beam intensity at a fixed experimental geometry.

Equation (6.6) also reveals that $I_{\text{SFG}}(\omega_3)$ is proportional to $\left|\chi_{\text{eff}}^{(2)}\right|^2$. In this expression, the second-order susceptibility tensor is replaced by $\chi_{\text{eff}}^{(2)}$, a scalar quantity. The quantity $\chi_{\text{eff}}^{(2)}$ reflects a specific component of the overall response that is active in a particular experimental setup. Dependence of SHG intensity on the laser power is shown in Figure 6.2(B). Since power is proportional to intensity (see eq. (4.19)), the intensity varies quadratically with laser power. The larger the $\chi_{\text{eff}}^{(2)}$, the stronger the signal. This highlights a method by which second-order susceptibility terms are measured. Analyzing the intensity of the generated signal in different polarization configurations and comparing the ratios to known standards, components of the second-order susceptibility tensor can be experimentally determined.

6.2.3 Molecular Hyperpolarizability

Each element of $\chi_{ijk}^{(2)}$ is a macroscopic quantity arising from the collective, average behavior of numerous individual molecules (N), characterized by its own molecular polarizability. Molecular polarizability describes the response of the electron cloud of a single molecule or atom under an applied external electric field. Mathematically, $\chi_{ijk}^{(2)}$ is related to the molecular hyperpolarizability tensor $\beta_{lmn}^{(2)}$ as follows:

$$\chi_{ijk}^{(2)} = \frac{N_S}{\varepsilon_0} \langle R_{il}(\theta) R_{jm}(\phi) R_{kn}(\psi) \rangle \beta_{lmn}^{(2)} \tag{6.7}$$

Here, $\beta_{lmn}^{(2)}$ is known as the first-order molecular hyperpolarizability (or second-order molecular polarizability) and N_s is the number density of the molecules undergoing nonlinear interaction. Combining eq. (6.7) with eqs. (6.4) and (6.5), it is apparent that the coherent SFG intensity will vary quadratically with the number density of the molecule as shown in eq. (6.8). In other words, if the orientation of the adsorbed molecules is invariant with number density, the sensitivity of I_{SFG} is directly proportional to N_s^2.

$$I_{SFG} \propto N_s^2 \left| \beta_{eff}^{(2)} \right|^2 \tag{6.8}$$

Similar to the second-order susceptibility, $\beta_{lmn}^{(2)}$ is also a third-rank tensor. They are connected via Euler transformation, which is expressed as a product of three rotational matrices $\langle R_{il}(\theta) R_{jm}(\phi) R_{kn}(\psi) \rangle$. We can get a pictorial description of the transformation from Figure 6.3, which shows two laser fields interacting with a molecular dipole at a planar surface to generate an SFG field. The rotational matrices connect the laboratory coordinate (x, y, z) and the molecular coordinate (x', y', z') systems using all three Euler angles (θ, ϕ, ψ). The angular brackets signify an ensemble average over the molecular orientational distribution. The indices i, j, k and l, m, n in eq. (6.7) denotes the unit vectors along the laboratory and molecular coordinates, respectively.

In general, using different polarization combinations of the fundamental (γ_{ω_1} and γ_{ω_2}) and the detected SFG fields ($\gamma_{\omega_1+\omega_2}$), the nonzero $\chi_{ijk}^{(2)}$ elements can be obtained. By virtue of eq. (6.7), the ratio of any $\chi_{ijk}^{(2)}$ element is related to the ratio of the angular average of the molecular polarizability tensor elements $\langle \beta_{lmn}^{(2)} \rangle$ projected in different directions. This connection involves a quantity known as the orientation parameter, defined as $D = \frac{\langle \cos^3\theta \rangle}{\langle \cos\theta \rangle}$, where θ is the molecular tilt angle – the angle between the molecular symmetry axis and the surface normal. Given this relationship, it is possible to determine the relative orientation of the molecule from the measured $\chi_{ijk}^{(2)}$ terms. This approach of determining molecular orientation has been elaborated and applied for various planar interfaces [94]. These calculations also involve the knowledge of experimental geometry, Fresnel coefficients, and molecular symmetry, and certain approximations. The absolute orientation, however, can be determined if the phase of the generated second harmonic (SH) or sum frequency signal is known. This is possible by means of phase-sensitive electronic or vibrational SFG

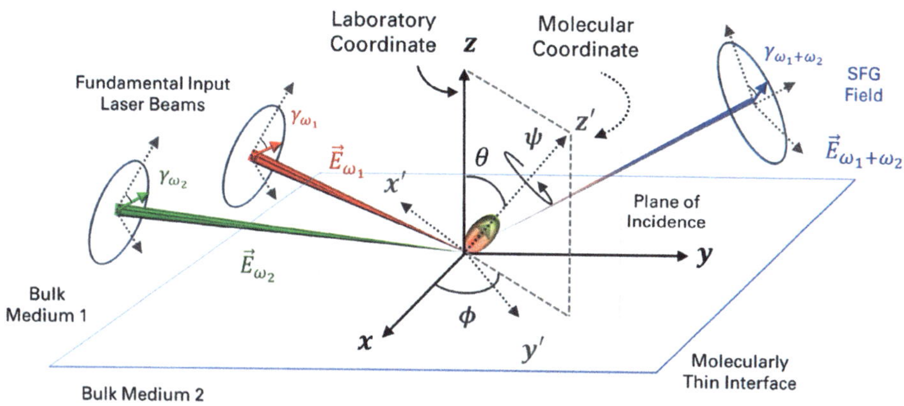

Figure 6.3: Sum frequency generation from a planar interface and the relationship between the laboratory and the molecular geometry.

heterodyne detection (HD) techniques (a technique that measures the field instead of intensity). Polarization-dependent SFG measurements and phase information are also important in deciphering the structure of colloidal interfaces, as discussed in Section 6.3.

At this juncture, it is necessary to highlight a few other relationships between SFG response and molecular hyperpolarizability. In Chapter 4, we have seen the linear first-order susceptibility as a function of the frequency field (eq. (4.10)). Similarly, frequency dependence of $\beta_{ijk}^{(2)}$ is expected. The total hyperpolarizability can be broken into a nonresonant and a resonant component. The resonant part is related to the transition dipole moment and contains a frequency-dependent denominator that gives rise to Lorentzian-shaped peaks in the SFG spectrum. The general relationship is

$$\beta_{ijk}^{(2)}(\omega_3;\,\omega_1,\omega_2) = \frac{1}{\hbar^2}\sum_{v,v'}\frac{\mu_{gv'}^{i}\mu_{gv}^{j}\mu_{vv'}^{k}}{(\omega_3-\omega_{v'g}+i\Gamma_{v'g})(\omega_2-\omega_{v,v'}+i\Gamma_{v,v'})}+\cdots \tag{6.9}$$

Evident from this relationship is the fact that, when the frequencies of the laser fields (either ω_1 or ω_2) or that of the SFG field (ω_3) match the molecular transition frequency (ω_{eg}), the denominator in eq. (6.9) becomes very small, and the overall $\beta_{ijk}^{(2)}$ increases considerably. Recall, $\beta_{ijk}^{(2)}$ is proportional to $\chi_{ijk}^{(2)}$ (eq. (6.7)), which in turn is proportional to SFG intensity (eq. (6.8)). This leads to an enhanced SFG signal, commonly known as resonance enhancement. Figures 6.4(A) and 6.4(B) provide a comparison between the nonresonant and a resonant case, respectively. While the energy diagram depicts an overlap between ω_{eg} and ω_1 in Figure 6.4(B), it is also possible to match laser frequencies such that $\omega_{eg} = \omega_2$ and $\omega_{eg} = \omega_3$ are achieved. These scenarios will likewise produce resonant SFG signals. When the frequencies do not overlap, the nonresonant SFG signal is weak.

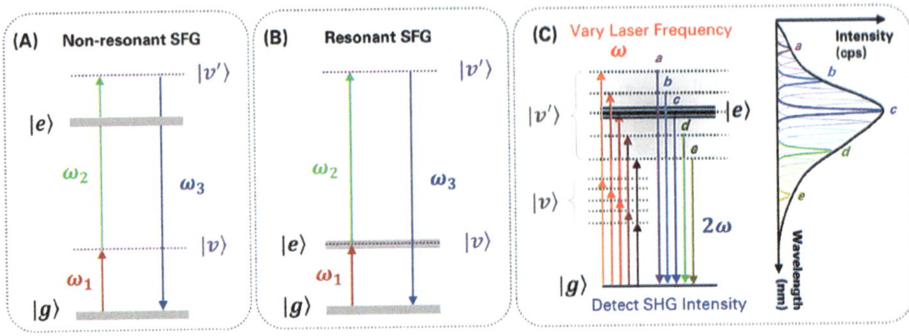

Figure 6.4: Energy diagrams describing the SFG process, highlighting the different (A) nonresonant and (B) resonance scenarios, in which $\omega_{eg} = \omega_1$. Electronic and virtual states are represented as solid and dashed lines, respectively. In the nonresonant case, the dashed and the solid lines do not overlap. (C) Energy transitions showing the process of surface SHG spectrum generation. Mapping of the energy landscape of the molecule is possible by tuning the frequency of the laser (as shown) or using a broadband laser. Transitions between energy levels are shown as arrows. The responses are shown as Lorentzian peaks.

Resonance enhancement is an important concept in spectroscopy for two reasons: (1) a more intense signal means a greater level of detectability as it is better distinguished from the background noise, and (2) it allows detection of specific molecular or vibrational transitions and thereby facilitates identification of a given molecule or functional group. By tuning the frequency of the fundamental laser beam or using a broadband laser, it is possible to scan the molecular transition frequency. In Figure 6.4(C), SHG spectrum generation is portrayed as an example of this process. When the laser frequencies fall in the visible/UV regime, electronic transitions are mapped out. SHG or electronic sum frequency generation serves the purpose of probing electronic transitions in molecules. It is possible to tune one of the incident laser frequencies to the IR region of the spectrum. In this case, SFG allows detection of vibrational modes in molecules. Recall, many of the stretches and bending modes of chemical bonds occur within 400–4,000 cm^{-1} (Figure 4.3). When SFG is used to probe vibrational modes, the technique is aptly called vibrational sum frequency generation or VSFG. Often, a broadband IR infrared pulse and a narrowband visible pulse are mixed to generate a broadband SFG spectrum, which provides information about vibrational modes in a single, simultaneous acquisition. Thus, this approach is faster and yields more detailed analysis compared to traditional scanning or tunable SFG techniques. The nonlinear process of DFG (see Section 6.1) is used in creating tunable mid-IR laser beams.

When applying VSFG to probe vibrational modes, the following important relationship needs to be considered: $\beta^{(2)}_{ijk,q} \propto \frac{\partial \alpha_{ij}}{\partial Q_q}\frac{\partial \mu_k}{\partial Q_q}$. Respectively, the partial derivatives describe the change in the Raman polarizability tensor component (α_{ij}) and the IR transition dipole moment of the qth vibrational mode with respect to the normal coordinate (Q_q) of the same mode. If either of these two quantities is zero, then no VSFG signal will be observed. In other words, the molecular polarizability and the transition dipole moment must change during the vibration. Vibrational modes must be IR and Raman active for it to be observed using VSFG.

6.2.4 Significance of Sum Frequency Generation

NLS, which includes many other second- and higher-order processes in addition to SFG discussed herein, has broad applications. SFG is a broad terminology of more specific processes that include SHG, VSFG, and electronic SFG. Its applicability spans material science, optical communications, crystal identification, medical diagnostics, drug development, environmental chemistry, and, as is the focus of this book, colloidal and interface science. Within the realm of interfacial chemistry, SFG is noteworthy because it is a highly surface-sensitive spectroscopic technique. It serves as an in situ and real-time probe for a wide range of buried interfaces that exhibit molecular thickness. It can be and has been applied to solid-solid, solid-gas, liquid-gas, solid-liquid, and liquid-liquid interfaces. The use of lasers with short pulses also provides temporal resolution to study ultrafast molecular events at various chemical interfaces.

Here is a recap of the fundamental essence and pertinent information these techniques can provide: VSFG allows acquisition of vibrational spectra, thus interfacial chemical composition is obtained. Since SFG is sensitive to an anisotropic environment, it can distinguish between ordered and random structures. Thus, it is possible to ascertain structures and orientations of proteins, peptides, hydrocarbons, and other molecules at buried interfaces. SFG can determine the average tilt angle and orientational distribution of functional groups relative to the surface normal by using different polarization combinations for the input laser beams. Likewise, SFG serves as a powerful tool to decipher interfacial water structure and hydrogen bonding. By the use of circularly polarized laser light, the inherent chirality of interfacial species can be obtained using SFG. Given that SFG intensity is proportional to the number density of the interfacial bound molecules, information about surface coverage, as well as the energetics of adsorption, is obtainable using these techniques. Real-time evolution of molecular species at interfaces can be probed. This allows for the study of adsorption/desorption kinetics, reaction kinetics, reorientation of molecules upon environmental changes, and ultrafast dynamics. Additionally, SHG/SFG is used to study charge properties of interfaces. Indeed, these techniques have been applied to study colloidal interfaces, a topic that we turn to next.

6.3 Sum Frequency Generation from Colloids: Fundamentals

Within recent decades, applications of SFG, including VSFG, pertinent to colloids have surged. Excellent reviews on this topic are available for curious minds interested in digging deeper [95–99]. Historically, the earliest demonstrations of second-order nonlinear phenomena, including interfacial effects, often involved SHG. This is because SFG experimental setups (see Section 6.3.2) are generally more complex than those of SHG. The use of colinear beams and a single frequency as the two fundamental fields makes SHG relatively facile. SH light emanating from molecules adsorbed at the colloidal/aqueous interfaces was first reported [100] in 1996. A 100 fs Ti:sapphire laser with the fundamental laser wavelength set at 854 nm was used. Scattered SH light from micron-sized centrosymmetric polystyrene (PS) latex beads coated with an organic dye, malachite green, was detected at 427 nm. The scattered SH intensity was demonstrated to depend quadratically (see eq. (6.8)) on the adsorbed species, a hallmark of the coherent second-order process. However, other processes can yield light at the SH (or SFG) frequency. This is particularly true when scattered light from a colloidal suspension is considered. It is important to be able to distinguish these sources.

6.3.1 Incoherent Versus Coherent Light Scattering

Figure 6.5 depicts the general scheme of light-particle interaction for the nonlinear scattering process. The diagram highlights many of the important parameters, which are discussed below. The second-order susceptibility, $\chi^{(2)}$, is one source of the scattered SF signal. The origin of the signal at the sum frequency from $\chi^{(3)}$ is discussed in Section 6.4. SFG is inherently a coherent process. SFG originating from anisotropic material, such as chemical interfaces or anisotropic crystals, involves phase-matching. The input fields must be spatially and temporally overlapped, and the resulting SFG photons originating from the same source are in phase. That is, the SFG photons maintain the same temporal and spatial coherence as those of the original fields. The intensity of the coherent response, as described in eq. (6.8), scales quadratically with the molecular concentration or number density. From here on, the sum frequency and SH scattering from colloids will be termed as SFS and SHS, respectively.

In SFS experiments, the fundamental and the SF fields are not parallel to each other (see Figure 6.5). The scattering wavevector is $\vec{q} = \vec{k}_0 - \vec{k}_3$, where $\vec{k}_0$ is the SFS wavevector and $\vec{k}_3$ is the sum of the wavevectors corresponding to the fundamental beams. As in eq. (5.1), the magnitude of $\vec{q}$ depends on the scattering angle as follows: $|q| = 2k_0 \sin\left(\frac{\theta}{2}\right)$. This phase relationship is important as the size of the particle becomes a factor in dictating interference in SFS (see Section 6.3.4). In describing SFS from colloidal interfaces, a concept of an effective particle susceptibility, $\Gamma_{ijk}^{(2)}$, has been introduced [101]. It is related to the second-order susceptibility $\chi_{ijk}^{(2)}$ and takes into account

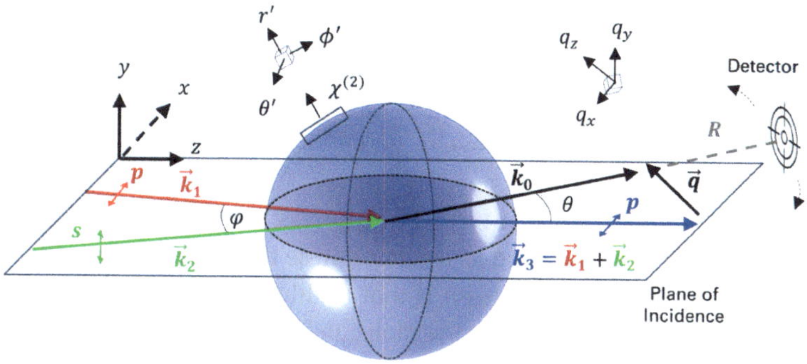

Figure 6.5: Schematic of nonlinear scattering from spherical colloids.

of the beam geometry and the symmetry prcperties of the adsorbate. Accordingly, an expression for the SFS intensity detected at distance R is

$$I_{\mathrm{SFS}} \propto \frac{1}{R^2}\left|\Gamma^{(2)} F\left[(\theta, \phi, d), \chi^{(2)}\right]\right|^2 I(\omega_1) I(\omega_2) \tag{6.10}$$

Here, $F\left[(\theta, \phi, d), \chi^{(2)}\right]$ is the form factor, which is a function of the scattering (in-plane) angle (θ), the out-of-plane angle (ϕ), and the particle diameter (d). Recall, the surface $\chi^{(2)}$ is related to the number density, and molecular hyperpolarizability (eq. (6.7)) connected through the tilt angle of the molecule on the surface. Thus, it is plausible to determine the molecular orientation of the adsorbate at the particle interface. As in the case of planar interfaces (Section 6.2.3), the molecular hyperpolarizability is related to electronic or vibrational transitions within the molecule and is frequency-dependent (eq. (6.9)). The SFS spectra thus yield absorption peaks corresponding to specific transitions. The spectra can be fit using a Lorentzian sum capturing the vibrational modes:

$$I_{\mathrm{SFS}}(\omega, \theta) \propto \left|\sum_n \left(A_{\mathrm{NR}} h(\omega, \theta) e^{i\Delta\phi} + \frac{A_n(\theta)}{(\omega - \omega_0) + i\Upsilon_n}\right)\right|^2 \tag{6.11}$$

The overall signal is composed of angle-dependent resonant ($A_n(\theta)$) contributions from the nth vibrational modes, and weak nonresonant (A_{NR}) amplitude. The term $h(\omega, \theta)$ describes the spectral shape of the dispersive background, $\Delta\phi$ is the phase difference between the signals, and Υ_n is the half-width at half maximum of the vibrational mode n.

Equation (6.11) describes the coherent SFS response from molecules from colloidal anisotropic interfaces. Signal at twice or the sum of the excitation frequencies can also originate from dipole induced isotropic medium by a nonlinear interaction of light. This is a result of local density fluctuations in scattering objects. As a result, the light emanating via this mechanism will be spontaneous, de-phased, and thereby incoherent. Hyper-Rayleigh scattering (HRS) and hyper-Raman scattering are two prime examples of the incoherent nonlinear elastic and inelastic scattering, respectively. The HRS intensity at 2ω is given as

$$I_{\mathrm{HRS}}(2\omega) = \frac{32\pi^2}{c\varepsilon_0^3\lambda^4R^2}F\sum_i N_i\langle\beta_{\mathrm{HRS},i}^2\rangle I^2(\omega) \tag{6.12}$$

Given that there can be multiple different types of scatterers, such as the solute and solvent molecules, the intensity is provided as a sum of these ith species. A key difference from the coherent SFS intensity (eq. (6.8)) is that in dilute solutions, HRS varies linearly with the concentration of the scattering molecules (eq. (6.12)). Like SHS, this intensity also scales quadratically with the intensity of the fundamental laser, and its source β_{HRS} depends on several factors, including scattering geometry, polarization state of the fundamental and the SH light, and molecular symmetry. Comparing this relationship with eq. (4.11) shows similarities and differences with Rayleigh scattering. Molecular electronic polarizability is the source of Rayleigh scattering, which is linearly dependent on laser intensity. A couple of major similarities are that HRS intensity is also inversely proportional to the fourth power of the fundamental wavelength and to the square of the distance of the scattering molecule. Instrumental parameters and collection optics that influence the efficacy of the I_{HRS} are described by the proportionality constant F. This relationship shows that HRS can be used to measure molecular hyperpolarizabilities. Indeed, this technique has been used to characterize and develop materials for efficient NLO response. These materials not only include organic chromophores but also colloidal substances [96].

Recall, SFS (SHS) is interfacial selective. However, signals due to HRS can be mixed in the overall detected signal from the colloidal suspension. Experiments often require subtraction of HRS background intensity from the solvent to thoroughly analyze the SFS/SHS intensity from the colloidal interface. The difference in the scaling with molecular number density is one way to distinguish between these processes. Other methods, such as temporal and spectral separation, are useful for distinguishing SFS from hyper-Raman and two-photon emissions, which are also incoherent. For light in the visible range, HRS is limited to molecules and particles no larger than 5–10 nm [97].

6.3.2 Experimental Aspects of SFS, SHS, and HRS

Ultrafast pulsed lasers are used in performing nonlinear scattering experiments from colloids. For instance, a femtosecond Ti:sapphire oscillator with a repetition rate in the order of MHz is common for SHS experiments. Figure 6.6 highlights different types of experimental setups involved. All these experiments involve the basic ingredients of (1) modulation and attenuation of the fundamental laser beam(s), (2) focusing and collimation of the light before and after the sample, (3) spectral separation and detection of the generated light, and (4) acquisition of data. Mirrors are used to guide the laser beams. Half -or quarter-waveplates and polarizers are used to modulate and select specific polarizations of light. For example, polarizations for the fundamental beams can be selected to be perpendicular to the plane of incidence (or *s*-polarized) and the scattered light, parallel to this plane (or *p*-polarized). In this scenario, the scattered intensity is labeled as I_{pss}. Figure 6.5 shows the orientations of the polarizations with respect to the plane of incidence. Combinations of a polarizer and a half-waveplate, or neutral density filters, can be used to attenuate the power of the laser beam.

The fundamental beam is then focused onto the sample using a lens. Prior to focusing the light onto the sample, a bandpass filter centered around the fundamental wavelength is used to eliminate any spurious signal at 2ω that may have originated from the optical components. In most SHS and HRS experimental setups, the fundamental fields are from the same light source and are focused on the sample collinearly (Figure 6.6(A),(B)). In SFS experiments, the two separate fundamental beams are brought in at an angle, as shown in Figure 6.6(C). They must be spatially and temporally overlapped at the focal point to generate the SFG signal. There are various ways to present the colloidal sample to the light. It can be introduced as a stream of liquid jet driven via a flow pump system, as shown in Figure 6.6(A). The sample can also be contained in a cell or cuvette with windows transparent to the fundamental and SFS wavelengths. Cells made of CaF_2 material are often used for VSFS experiments that use IR light. Colloidal suspension can be flowed through the cuvette or stirred using a magnetic stirrer. Additionally, a temperature-controlled cuvette holder can be used in this setup. As discussed below, the thickness of the cell, along with the particle number density, affects the efficacy of the signal. Moreover, interfacial regions of the cell wall and the fluid can be a source of SFG; thus, best practice involves verifying the origin of the signal and performing control experiments.

The scattered light is collimated. The collection half angle of the scattered light can be calculated using $\theta = \tan^{-1}\left(\frac{D}{2f}\right)$, where f and D correspond to the focal length and the diameter of the lens. As an example, the collection angle using a lens of $f = 10$ cm and $D = 2.5$ cm is approximately 14°. Thus, to obtain angular resolved SFS data, a pinhole or an iris is used to limit the solid angle of detection (see Figure 6.6 (A)). Use of a band-pass filter centered around the SFG wavelength blocks the residual fundamental beam. A polarizer is used to select the desired polarization of the SFS light. Collimated light is then sent to the monochromator for further spectral separation. For angle-resolved SFS/SHS experiments, optical fibers are used (Figure 6.6(A)), which allow flexibility in rotating the detection arm using a goniometer. The detector rotates, allowing the collection of the scattered light at a wide range of angles relative to the incident laser beam. For experiments not requiring angular resolution, the detecting angle is fixed, and an appropriate lens is used to gather most of the scattered light. In an HRS experiment, the light is generally collected at a 90° angle (Figure 6.6 (B)). For the collection of light, photomultiplier tube or charged-coupled device cameras are typically used.

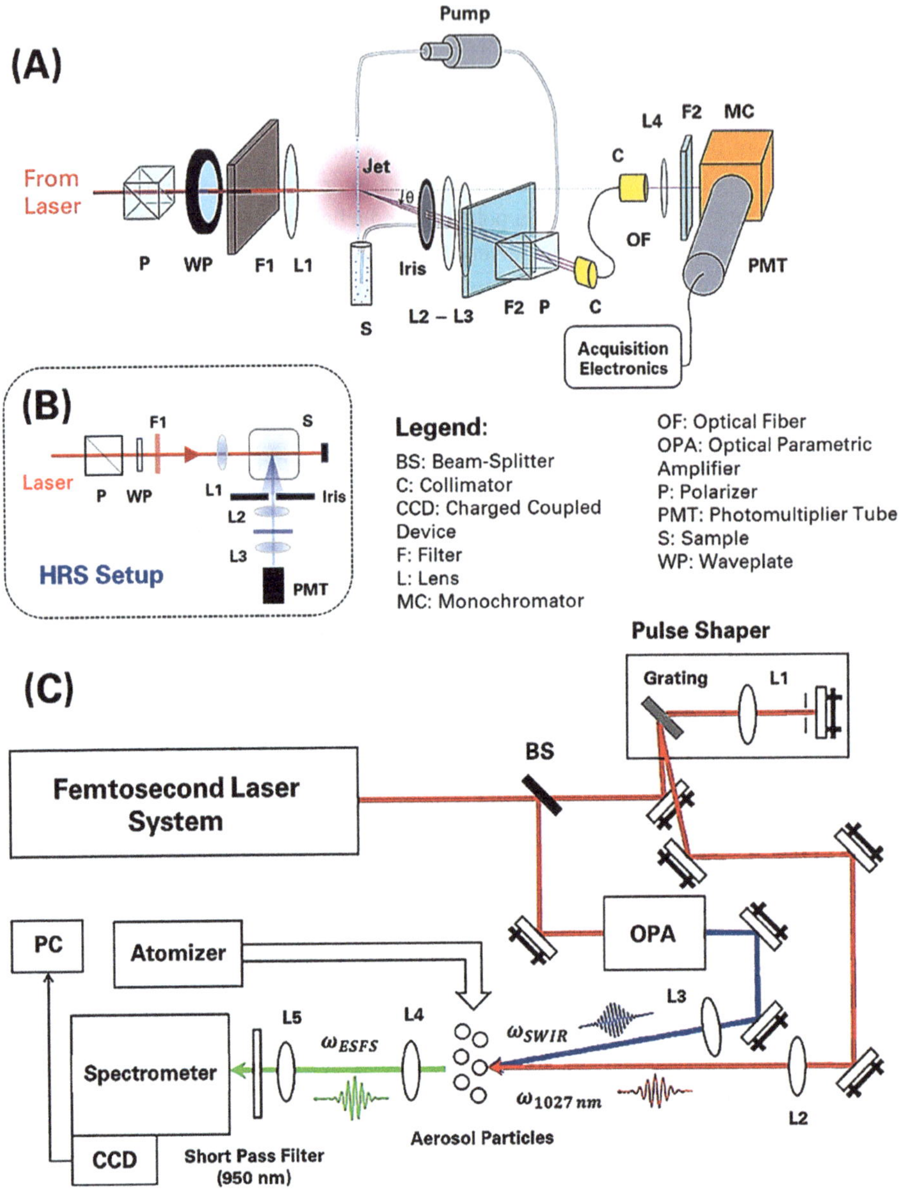

Figure 6.6: Schematic of nonlinear scattering from colloids, highlighting key experimental parameters. (A) Typical SHS setup with the angle-resolved detection. Adapted and reprinted with permission from Gonella and Dai [98]. Copyright 2014, American Chemical Society. (B) HRS collected at 90°. (C) Experimental setup for ESFS detected in the forward direction. Adapted and reprinted with permission from Qian et al. [102]. Copyright 2020, American Chemical Society.

Special care must be considered when performing SFS experiments with IR and visible beams. SFS experimental setup requires noncollinear geometry and careful temporal and spatial overlaps of the beam. Proper optics that are transparent to IR wavelengths are also used. Moreover, water molecules exhibit strong absorption of IR light. Thus, D_2O is often used to prepare a colloidal suspension. Heavy water has a smaller IR absorption cross-section than H_2O. The laser beam is focused on a smaller volume of the sample to avoid significant absorbance of light. In summary, the efficiency of SFS/SHS depends on various factors. In eq. (6.11), the dependency on the material system and the laser intensities are highlighted. With respect to experimental components, the number of photons generated is also directly proportional to the pulse duration, the repetition rate of the laser, the focal area of the beam, and the detector efficiency.

6.3.3 Optimized SHS Signal from Colloids

The amount of scattered sum frequency and SH light detected from a colloidal suspension is dependent on the particle density and the pathlength of the sample. Assuming full surface coverage, increasing the particle number density can be expected to increase the SHS signal according to eq. (6.8). However, this happens to a certain extent. If the particle density is too large, absorption and multiple scattering of the generated signal diminish the amount of light detected. This phenomenon has been explored using latex particles coated with malachite green, an SHG dye. Particles of various diameters in the range of 0.1–2.9 μm have been used [103]. SHG intensity was collected at a forward direction with an acceptance angle of 60°. For a fixed pathlength (l), there exists a particle number density at which a maximum intensity is observed. Below and above that number density, the signal is reduced. This work also provides an analytical formula (eq. (6.13)) describing the SHS intensity as a function of particle number density (N). This is an approximation that neglects the laser beam waist variation that can occur due to tight focusing. The exact solution requires integration over the path length that includes the variation of the focused beam profile. In this equation, α_ω and $\alpha_{2\omega}$ represent the extinction coefficients at ω and 2ω, respectively. Figure 6.7(A) shows the variation of SHS intensity for spherical particles of three different diameters:

$$I_{\mathrm{SHS}} = \frac{\beta_{\mathrm{eff}}^2}{2\alpha_\omega - \alpha_{2\omega}}\left(e^{-\alpha_{2\omega}Nl} - e^{-2\alpha_\omega Nl}\right) \tag{6.13}$$

Even though eq. (6.13) is an approximation, it predicts the qualitative feature of the SHS intensity profile quite well. At higher number densities, eq. (6.13) deviates from the exact solution for a tighter focal length of 10 cm or less. Figure 6.7(A) shows that

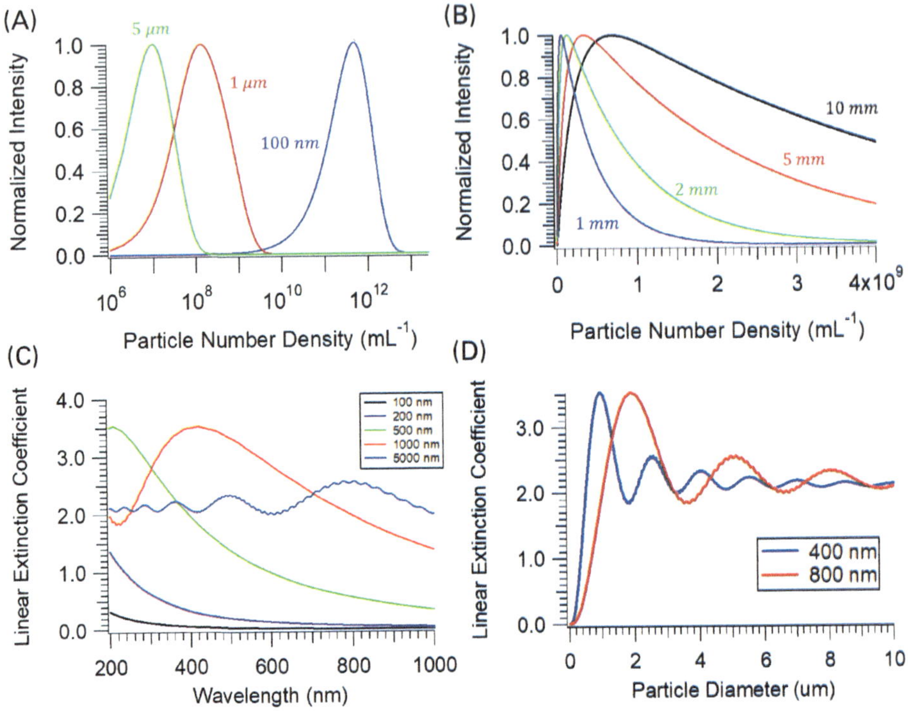

Figure 6.7: Simulation of eq. (6.13) describing SHS intensity from colloids as a function of particle density. (A) Effect of particle size with a fixed path length of 5 mm. (B) Effect of path length for a particle diameter of 1 um. In these simulations, extinction coefficients have been calculated based on Mie theory with the refractive index set at 1.5 + 0.05*i*. Calculated linear extinction coefficients based on Mie theory as a function of (C) wavelength and (D) particle size. In these calculations, the medium refractive index is 1.33, and that of the particle is 1.6 + 0.01*i*. Calculations have been performed using MiePlot v4621, an online computer program for scattering of light from a sphere [104].

there is a particle concentration at which a maximum occurs. By taking the derivative of eq. (6.13) and setting it equal to zero, we find the optimum particle density:

$$N_{\text{opt}} = \frac{1}{(\alpha_{2\omega} - 2\alpha_{\omega})l} ln\left(\frac{\alpha_{2\omega}}{2\alpha_{\omega}}\right) \tag{6.14}$$

The optimal number density depends only on the linear extinction coefficients and the optical path length of the cuvette. The effect of path length is shown in Figure 6.7 (B). The linear extinction coefficients can be measured experimentally using a UV-Vis spectrometer or calculated based on Mie theory. The geometrical setup for a UV-Vis measurement is generally not like that of the SHS experiment. The latter typically involves the collection of a large fraction of scattered light, resulting in a lower value of the extinction factor compared to the linear extinction coefficients.

Figure 6.7(A),(B) shows normalized SHS intensity to highlight the positioning of the optimum particle density. However, it is important to recognize that the efficiency of the SH intensity varies with particle size. SHS efficiency is defined as the amount of SH signal generated per surface-adsorbed dye molecule. A major factor that influences this efficiency is the loss of the fundamental and the SH light intensity due to Mie scattering. The scattering efficiency fluctuates with wavelength and particle size. Simulations of extinction coefficients as a function of wavelength for particles of different sizes are shown in Figure 6.7(C). In Figure 6.7(D), scattering extinctions for two distinct wavelengths are plotted against particle size. The extinction coefficients dictate how much SHS intensity is detected. Another factor, as noted in Chapter 4, is that light scattering also exhibits angular dependency. The SHS efficiency of PS particle surface coated with malachite green has been determined in the forward direction, using a fundamental wavelength of 840 nm. It has been shown [105] that SHS efficiency increases with particle size diameter up to approximately 1 um, where a maximum occurs. Thereafter, SHS efficiency decreases slowly with increasing particle size. The angular dependence of SHS intensity with particle size is presented in Section 6.3.5.

6.3.4 Phase Difference and Importance of Particle Size on SFS

Another aspect that influences the overall SHS signal is the phase-difference between the SHG fields from individual molecules separated by a certain distance. As the fields emanating from molecules located at different positions on the particle surface are added, they are subject to constructive or destructive interference. For simplicity, let us consider two molecules adsorbed at two different locations (s_1 and s_2) of a spherical particle. Two different scenarios, small versus large particles in relation to the wavelength of the light, are illustrated in Figure 6.8(A). Assuming the two molecules are oppositely oriented and that their hyperpolarizabilities are equal in magnitude but opposite in sign, the SHG field from the two dipoles can be written as

$$E_{\text{SHG}} \propto \left(\beta_{s_1}^{(2)} + \beta_{s_2}^{(2)} e^{i\vec{q}\cdot\vec{d}}\right) E_\omega E_\omega = \beta^{(2)} \left(1 - e^{i\vec{q}\cdot\vec{d}}\right) E_\omega E_\omega$$

The quantity $\vec{q}$ is the nonlinear scattering vector (Figure 6.5) and $\vec{d}$ is the distance by which the two dipoles are separated, or in this diagram, the diameter of the particle. We see that as the separation is diminished to zero (or as the particle diameter decreases), the $e^{i\vec{q}\cdot\vec{d}}$ factor approaches 1. Consequently, E_{SHG} vanishes. When $\vec{q}\cdot\vec{d} = \pi$, a maximum constructive interference occurs. Note, $e^{i\pi} = -1$. In this context, d is representative of the coherence length, the distance over which the fundamental and the SH waves can remain in phase. In Section 6.2.3, we have seen that phase matching condition involves the configurations of the light beams and the refractive index at each wavelength. The implication of the product $\vec{q}\cdot\vec{d}$ reveals that for scattering from colloids, the overall SHS or SFS field depends also on the size, shape, and structure of the particle. Dye molecules adsorbed at isotropic particles of a diameter in the order of sub-100 nm yield a weak or negligible SHG signal in the forward direction. However, significant SHS in the forward direction can be observed from larger particles. This phase-relationship property has a significant implication in observing molecular transport across biological cell membranes, the thickness of which is, in general, a few nanometers (Chapter 7).

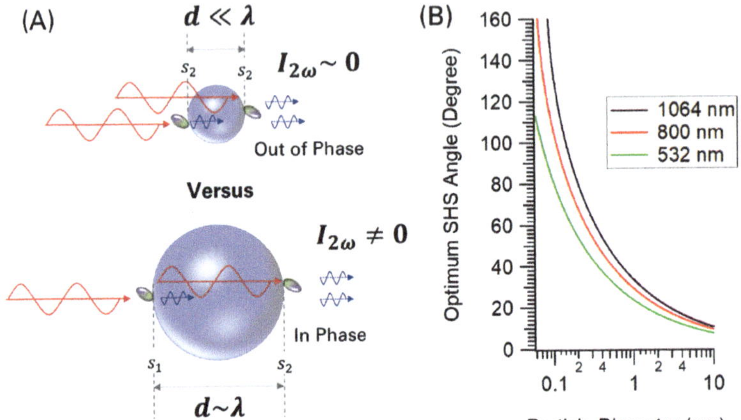

Figure 6.8: (A) Effect of particle size on the phase relationship of SHG fields. (B) Optimum SHS angle as a function of particle size at different laser wavelengths.

6.3.5 Angle-Resolved SHS

The following phase matching condition for constructive interference applies [105] for SHS: $\vec{q} \cdot \vec{d} = \frac{4\pi n_{2\omega}}{\lambda_{2\omega}} d \cdot \sin^2\left(\frac{\theta}{2}\right) = \pi$. Solving for the scattering angle, we obtain

$$\theta = 2 sin^{-1} \sqrt{\left(\frac{\lambda_{2\omega}}{4 n_{2\omega} d}\right)} \tag{6.15}$$

where θ is the in-plane angle (Figure 6.6). The SHS wavelength and the refractive index at 2ω are represented as $\lambda_{2\omega}$ and $n_{2\omega}$, respectively. Equation (6.15) reveals that the phase-matched SH intensity tilts toward a smaller scattering angle with increasing particle size. This is demonstrated in Figure 6.8(B) for three different laser wavelengths. Equation (6.15) provides a qualitative insight into how particle size and refractive index can impact the angle at which the SHS intensity is dominant. In 2001, the first AR-SHS measurement on centrosymmetric particles was performed using an aqueous-based colloidal suspension of malachite green adsorbed onto micron-sized PS beads [106]. The first angle-resolved sum frequency scattering (AR-SFS) from SiO_2 particles dispersed in CCl_4 was reported in 2003. Subsequent research provided a further theoretical basis for the angular-dependent scattering pattern and the interfacial information possible from these measurements.

The intricate scattering pattern can be explained based on RGD and Mie theories and their extensions to nonlinear scattering [107–109]. AR-SHS experiments using dye-coated PS particles have been conducted to validate these models. The nonlinear RGD model is applicable to small particles, with diameters up to approximately 200 nm. Nonlinear Mie theory is applicable to a wider range of particles. It has been shown that: (1) small particles with diameters less than 100 nm scatter light with a single maximum around 90° relative to the direction of the fundamental laser beam, (2) the AR-SHS pattern for larger particles (diameter $\geq 1\ \mu$m) is prominent in the forward direction and features multiple oscillations. Figure 6.9(A) provides examples of AR-SHS data for particles of different sizes and fundamental/SHS polarization combinations. It is apparent that molecules adsorbed on the surface of the micrometer-sized particles can be probed in the forward direction, and nanometer-sized particles at larger angles.

The variation in the angle-resolved scattering pattern is not only particle size dependent but also strongly sensitive to the polarization states of the incident and detected light. We can visualize this by comparing the top panel (PP polarization combination) and the bottom panel (SP polarization combination) in Figure 6.9(A). Recall Section 6.3.2; the ratio of these intensities is related to the ratio of susceptibility elements, which in turn is coupled to the average tilt angle or the average orientation of the molecule on the surface. In the case of planar interfaces, reflection geometry allows a direct connection between molecular hyperpolarizability and the surface susceptibility via average orientation. In the scattering experiment, the effective particle susceptibility, which is dependent on the scattering geometry and particle size, needs to be considered. Nevertheless, the adsorbate orientational information is embedded in the angle-resolved scattering data.

To illustrate this concept, simulations of scattering patterns for the PP and SP polarization combinations are reproduced [110] in Figure 6.9(B). This simulation considers water molecules (C_{2v} point group) around a particle of 1 μm diameter, and the fundamental wavelength is set at 800 nm. It is apparent that at low scattering angles, intensity variation is negligible with tilt angles. At higher scattering angles, the overall signal decreases. Thus, an interval of 14°–30° has been suggested as an optimal scattering range for orientational analysis based on AR-SHS. The angular range corresponds to the qR range of 1.4–4, where R is the particle radius. The same ideas apply to AR-SFS experiments, where different vibrational modes are probed. It has been shown that due to intrinsic properties of the molecule, AR-SFS of certain stretches (e.g., symmetric stretches in H_2O and CH_2) are not sensitive to molecular tilt angle. The optimal AR-SFS scattering range is 17° to 50° ($1.4 < qR < 4$)for the asymmetric stretch mode of C_{2v} and C_{3v} point groups for the particle with a diameter of 1 μm. This optimal angle range changes with particle size. Figure 6.9(C) displays the optimal angular range versus particle size for both SHS and SFS experiments. The finding shows

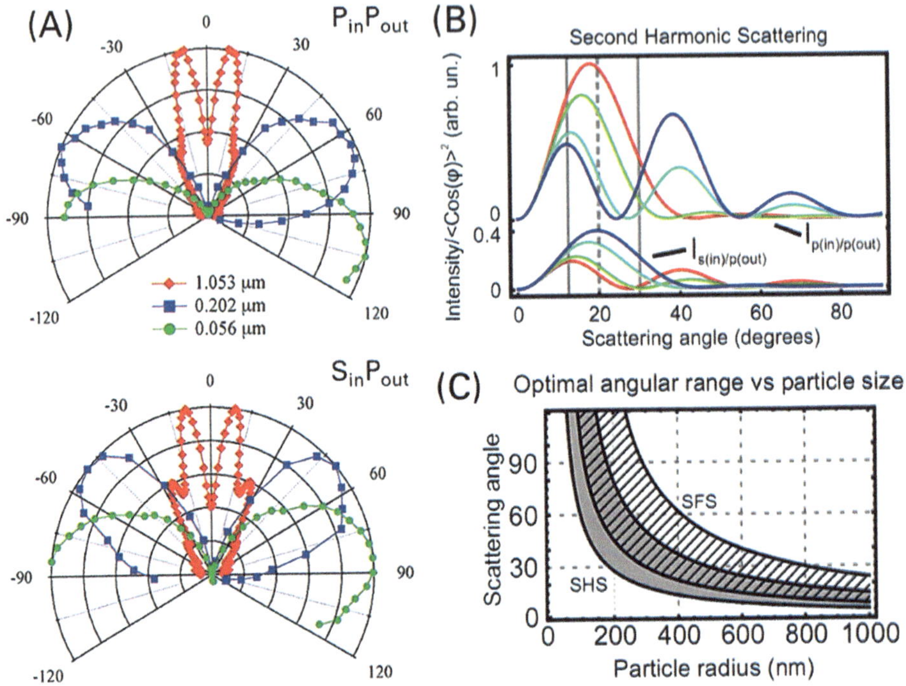

Figure 6.9: (A) Effect of particle size and polarization combinations on the AR-SHS pattern. Adapted and reprinted with permission from Jen et al. [105]. Copyright 2009, American Chemical Society. (B) AR-SHS intensity simulation for PP and SP polarization combinations as a function of scattering angle [110]. The effect of molecular orientation (tilt angle, ϕ) is represented with different colors: $\phi = 0^\circ$ ($D = 1$) and $\phi = 90^\circ$ ($D = 0$) are the red and the blue curves, respectively. The intermediate curves represent values of 30° ($D = 0.75$) and 60° ($D = 0.25$). (C) Angular position of the optimal interval of ($1.4 < qR < 4$) for SFS wavelength of 645 nm (hatched area) and SHS wavelength of 400 nm (gray area) [110]. The lower and upper curves each correspond to the boundaries of $qR = 1.4$ and $qR = 4$, respectively. The orientation parameter D determines the relative values of the $\chi^{(2)}$ components. Adapted and reprinted with permission from de Beer and Roke [110]. Copyright 2010, AIP Publishing.

that the larger the size of the particle, the smaller the scattering angle must be selected for molecular orientation determination. The optimal angular range is higher for smaller particles. The optimum range occurs at larger angles for SFS than for SHS.

6.4 $X^{(3)}$ Contribution to SFS Intensity

Our discussion so far involved sum frequency (SH) generation due to $\chi^{(2)}$ response of a material. There is another process by which SFS (or SHS) is generated, and it occurs when the material is in the presence of a DC field (E_{DC}). Referring to Table 6.1 and the preceding example, we see that DC field-induced SFG involves the third-order susceptibility properties of molecules. When we consider a charged interface in contact with the aqueous phase, the DC field is primarily due to the interfacial charge density. Be it a planar or colloidal interface, as shown in Figure 6.10(A), the field emanating aligns the water molecules in the vicinity of the electric double layer. In this schematic diagram, the negative end of the water molecules is pointing away from the negatively charged surface. The larger the surface charge density, the stronger the field and the greater the extent of polarization, which can reach several nanometers into the bulk. Although $\chi^{(3)}_{H_2O}$ is dominated by field-induced reorientation of water molecules, field-induced changes of the H-bonding structure and electronic polarizabilities of molecules can also contribute to the "Chi-3" phenomenon.

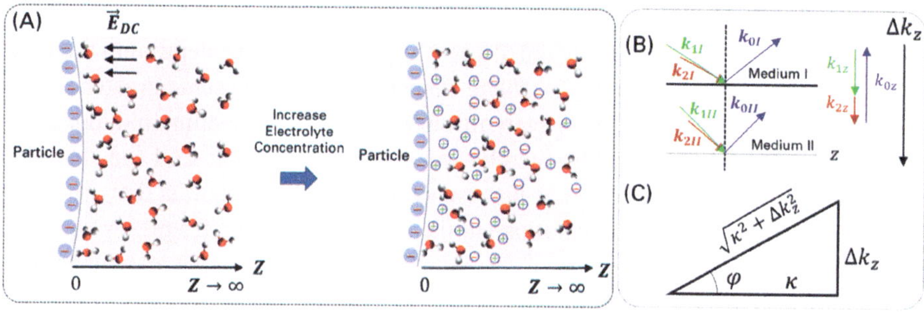

Figure 6.10: (A) Effect of electrolyte on the alignment of water molecules near a charged interface. (B) Vectorial depiction of the wavevector mismatch and the probing depth.

The $\chi^{(3)}_{H_2O}$ contribution to the second-order induced polarization, $P^{(2)}$, can be shown to be related to the surface potential, ψ_0 [111]. Based on the diagram, we can write

$$P^{(2)} = \chi^{(3)}_{H_2O} E^2_\omega \int_0^\infty E_{DC}(z)dz$$

Here, E_ω represents the fundamental field. Assuming $\chi^{(3)}_{H_2O}$ is constant over the distance from the surface into the bulk, and recognizing $E_{DC}(z)dz = -d\psi(z)$, the integration yields $-\chi^{(3)}_{H_2O}E^2_\omega[\psi(\infty) - \psi(0)]$ at a large distance from the surface $\psi(\infty) = 0$. Thus, we get $P^{(2)} = \chi^{(3)}_{H_2O}E^2_\omega\psi_0$. The overall SHG field from a charged interface, where both second- and third-order processes contribute, is thus written as

$$E_{2\omega} \propto \chi^{(2)} + \chi^{(3)}_{H_2O}\psi_0 \tag{6.16}$$

This relationship assumes that the optical fields are independent of z, and thereby the phases of the incoming and returning fields do not change in the presence of E_{DC}. Incorporating the Gouy-Chapman model (eq. (3.2)) results in eq. (6.17), where c_i is the electrolyte concentration:

$$E_{2\omega}(c_i) = \chi^{(2)} + \chi^{(3)}_{H_2O} \frac{2k_BT}{z_ie} \sinh^{-1}\left(\frac{\sigma_0}{\sqrt{8c_i\varepsilon N_A k_B T}}\right) \tag{6.17}$$

Thus, by measuring the SHS intensity as a function of electrolyte concentration, it is plausible to determine the interfacial charge density, σ_0, and the surface potential. This technique has been utilized to study charged planar, as well as colloidal interfaces (see Chapter 7).

However, it is important to understand the limitations of this approach of extracting surface potential and charge density from SFS/SHS measurements in only a forward direction from colloidal suspensions. There are several factors to consider. For instance, due to the presence of a DC field and reorientation of water molecules, a small amount of centrosymmetry is broken. This, in turn, changes the thickness of the interfacial region being probed by the SFS technique. The probing depth into the bulk solution, which depends on the electrolyte concentration, can become greater than 1 μm [112]. As the optical fields propagate through a dispersive medium and along the extended probing depth, interference between the optical fields and the spatially varying electrostatic fields occurs. A key aspect of this phenomenon is that a mixing of second- and third-order contributions to the overall SH or sum frequency signal is expected [112–115]. Taking account of this interference, and assuming an exponential decay of the electrostatic potential with distance z (Figure 6.10), an updated model describing the SH response is provided as follows [112]:

$$E_{2\omega} \propto \chi^{(2)} + \chi^{(3)}_{H_2O}\psi_0 f(k, \Delta k_z) \tag{6.18}$$

where the correction factor $f(k, \Delta k_z)$ is given as $\frac{\kappa}{\kappa - i\Delta k_z}$. In this expression, κ^{-1} is the Debye length (Section 3.2) and Δk_z is the wavevector mismatch for SHG/SFG fields generated at different probing depths. Figure 6.10(B) illustrates the probing depth and the definition of Δk_z. The correction factor, which corrects for the interference, can also be expressed as $\Delta k_z \cos(\varphi)e^{i\varphi}$, where φ is the phase angle (see Figure 6.10(C)) [114–116]. The phase angle is related to the Debye-Hückel screening length, λ_D, as follows: $\varphi_{DC} = \tan^{-1}(\Delta k_z \lambda_D)$. The phase angle reflects the z-dependence of $\chi^{(3)}$ signal as it is produced throughout an extended region between the interface and the bulk solution. It does not represent the phase of the overall signal, which can include phase variation in $\chi^{(3)}$ near or off-resonance. Recent advances [116] have shown that it is possible to separate the $\chi^{(2)}$ and $\chi^{(3)}$ contributions to the overall signal and explain surface-potential dependent SFG spectra from charged aqueous interface based on HD method, such as HD-SHG in reflection geometry for planar interfaces. The overall SFG intensity incorporating the $\chi^{(3)}$ contribution and the interference effect is

$$I_{\mathrm{SFG}} \propto \left| \chi^{(2)}_{\mathrm{NR}} + \sum_{\nu=1}^{n} \chi^{(2)}_{\mathrm{res},\,\nu} e^{i\gamma_\nu} + \frac{\kappa}{\sqrt{\kappa^2 + (\Delta k_z)^2}} e^{i\varphi} \chi^{(3)} \psi_0 \right|^2 \tag{6.19}$$

As in eq. (6.11), the nonresonant and resonant $\chi^{(2)}$ contributions are present, but for charged interfaces, $\chi^{(3)}$ contributions become important.

These considerations are also applicable to SFS or SHS measurements for the analysis of colloidal charged interfaces; however, geometric factors influence the scattering intensity and make the data interpretation more complex. As discussed in Section 6.3.5, Mie scattering form factors (or other approximations) must be incorporated to extract accurate physical meaning from scattered intensity. Thus, in the case of colloids, intensity versus electrolyte concentration experiments performed at a fixed detection angle and polarization combination may provide qualitative understanding, but not quantitatively meaningful results. Instead, angular-resolved SHS/SFS in conjunction with polarization-dependent studies are necessary. Polarization and angle-dependent scattering patterns as a function of electrolyte concentrations not only serve to separate surface versus electrostatic contributions but also yield a better understanding of the interfacial structure of water molecules and the electric double layer.

End-of-Chapter Problems

1. Define and describe the following terms and concepts: induced polarization, NLS, phase matching condition, second-order susceptibility, molecular hyperpolarizability, SFG, SHG, third harmonic generation, anisotropic, resonance enhancement, scattering wavevector, HRS, and “Chi-3 effect.”

2. What are the fundamental advantages and limitations of using SHG spectroscopy versus other surface characterization techniques (e.g., TEM, DLS, zeta potential measurements, and extinction spectroscopy) for studying colloidal systems?

3. VSFS selection rule:
   (A) What is the primary symmetry requirement for a medium to generate a sum frequency signal via the electric dipole mechanism?
   (B) What are the molecular-level requirements for a specific vibrational mode of a molecule at the colloid interface to be observed in a VSFS spectrum?
   (C) How does molecular orientation at the interface affect the VSFS signal, and what information can be extracted by controlling experimental parameters?

4. A high-intensity laser beam with an electric field strength of $E = 10^7$ V/m is incident on a noncentrosymmetric crystal (e.g., GaAs). The crystal has a first-order susceptibility $\chi^{(1)} = 11.9$ and an effective second-order susceptibility $\chi^{(2)} = 1 \times 10^{-10}$ m/V. Estimate the magnitudes of the linear polarization $P^{(1)}$, and the second-order nonlinear polarization, and compare their relative strengths.

5. A fundamental laser beam with a wavelength of 1,064 nm is incident on a nonlinear crystal. The refractive index of the crystal is $n_\omega = 1.55$ at the fundamental frequency and $n_{2\omega} = 1.55$ at the SH frequency:
   (A) Calculate the magnitude of the fundamental wavevector k_ω and the SH wavevector $k_{2\omega}$ inside the crystal.
   (B) Calculate the phase mismatch parameter Δk assuming collinear propagation
   (C) Calculate the coherence length L_c over which the generated SH signal cancels itself out due to the phase difference.

6. SFG wavelength and detection:
   (A) A 1,064 nm Nd:YAG laser beam is mixed with its own frequency-doubled light (SHG) to produce UV light via SFG. What is the wavelength of the resulting UV light (in nm)?
   (B) A SFG spectroscopy experiment is performed using a fixed-wavelength visible (VIS) laser at 532 nm. Calculate the required wavelengths for the tunable IR laser beam to probe the following molecular vibrational stretches:
   A typical C=O (carbonyl) stretch
   A typical C–H (alkane) stretch

7. A visible laser beam with a wavelength of 500 nm and an IR beam at 3,000 nm are incident on a planar surface from air ($n = 1$). The incident angle of the visible beam is $\theta_{vis} = 45^\circ$, and for the IR beam, it is $\theta_{vis} = 60^\circ$. Both beams are in the same plane of incidence. Calculate the output angle θ_{SFG} of the generated SFG beam relative to the surface normal.

8. An SHS experiment uses an incident laser with a fundamental wavelength of 800 nm in an aqueous solution. The colloidal particles have a diameter of 200 nm. The refractive index of the medium at the SH wavelength is 1.33. Estimate the expected scattering angle for the maximum SH signal.

9. In an AR-SHS experiment, the scattering patterns are measured from two different samples of dye-coated PS colloidal particles (diameters of 50 and 500 nm) suspended in an aqueous solution. The fundamental laser wavelength is well outside any linear absorption bands.

(A) Describe the possible patterns.

(B) Explain the physical reasons for the difference in the observed angular resolved SHG patterning between the smaller (50 nm) and larger (500 nm) particles.

10 Consider a surface-bound C–CH_3 functional group oriented at a direction with a rotation angle ψ, azimuthal angle ϕ, and tilt angle θ with respect to the lab-frame coordinate $(x,\ y,\ z)$ as shown. The $a,\ b,\ c$ represent the molecular coordinate for the methyl group [117]. For the asymmetric C–H stretch, it can be shown that:

i. $\chi^{(2)}_{yyz} = \chi^{(2)}_{xxz} = -N\beta^{(2)}_{caa}[\langle\cos(\theta)\rangle - \langle\cos^3(\theta)\rangle]$

ii. $\chi^{(2)}_{zxx} = \chi^{(2)}_{zyy} = \chi^{(2)}_{yzy} = \chi^{(2)}_{xzx} = -N\beta^{(2)}_{caa}\langle\cos^3(\theta)\rangle$

iii. $\chi^{(2)}_{zzz} = 2N\beta^{(2)}_{caa}[\langle\cos(\theta)\rangle - \langle\cos^3(\theta)\rangle]$

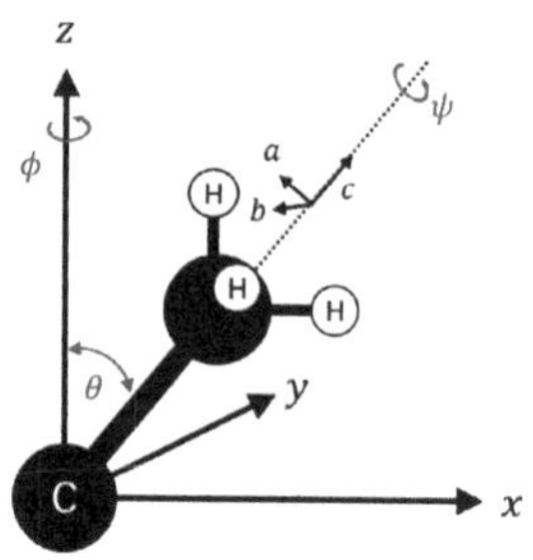

(A). Show that the ratio $\frac{\chi^{(2)}_{yyz}}{\chi^{(2)}_{yzy}} = \frac{\langle\cos(\theta)\rangle - \langle\cos^3(\theta)\rangle}{\langle\cos^3(\theta)\rangle}$

(B). Assuming the distribution of θ to be a δ-function, then $\langle\cos(\theta)\rangle$ and $\langle\cos^3(\theta)\rangle$ can be substituted by $\cos(\theta)$ and $\cos^3(\theta)$, respectively. If the experimental

ratio of $\frac{\chi^{(2)}_{ssp}}{\chi^{(2)}_{sps}} = \frac{\chi^{(2)}_{yyz}}{\chi^{(2)}_{yzy}} = 1.7$, determine the tilt angle.

11. How does the specific chemical functionalization (e.g., adding carboxyl groups) of the nanoparticle surface, and its resulting surface charge, affect the surrounding interfacial water orientation and the subsequent SHG signal?

12. Optimum particle concentration:
(A) Starting with eq. (6.13), derive the optimum particle density formula (eq. (6.14)).
(B) Calculate the optimum number density N_{opt} for a 700-nm colloidal solution used in SHG experiments, given the following parameters:
Absorption coefficient at $\lambda = 400$ nm $= 2.7 \times 10^{-9}$ mL/particle · cm
Absorption coefficient at $\lambda = 800$ nm $= 1.4 \times 10^{-9}$ mL/particle · cm
Path length of the sample = 1 cm.
[ans. 3.4×10^8 particle/mL]

13. A surface SFG experiment on water (a $\chi^{(3)}$ material at a charged interface) uses visible and IR beams to generate an SFG signal. Given the following parameters for a single resonance case ($n = 1$): nonresonant susceptibility $\chi^{(2)}_{NR} = 1$ pm/V, resonant susceptibility $\chi^{(2)}_{NR} = 5$ pm/V with phase $\gamma_1 = \pi/2$, coherence length term $\frac{\kappa}{\sqrt{\kappa^2 + (\Delta k_z)^2}} e^{i\varphi} \chi^{(3)} \psi_0 = 2e^{i\pi/4}$ pm/V. Calculate the magnitude of the total nonlinear effective susceptibility $|\chi^{NL}_{eff}|$.

14. In a reflection-geometry SFG experiment, the SFG intensity is proportional to the square of the effective susceptibility. The third-order contribution is significantly affected by the phase-matching condition Δk_z. Determine how the intensity changes when the wave vector mismatch Δk_z goes from a small value, where $(\Delta k_z)^2 \ll \kappa^2$, to a large value, where$(\Delta k_z)^2 \gg \kappa^2$, assuming all other terms remain constant.

7 Application of SFS and SHS: Interfacial Processes and Structure

We have learned the fundamental aspects of second-order nonlinear spectroscopy in Chapter 6. A major punch line is that SFG is an interfacial selective tool that provides a direct, label-free approach to probing various buried chemical interfaces. Many of the insights and advances in the technique stemmed from investigating planar interfaces, performing the experiment in reflection geometry. When applied to colloidal interfaces, sum frequency or second harmonic generation (SHG) involves the detection of scattered light. These SFS/SHS techniques, which include interfacial vibrational/electronic spectroscopy and angle-resolved studies, provide a wealth of information about processes that occur at colloidal interfaces. In addition, the interfacial structural properties of colloids can be elucidated using these methods. SFS/SHS have been applied to study colloids in the realm of biological, medical, atmospheric, environmental, and material sciences. In this chapter, some of the pioneering results and key findings pertinent to different types of colloids, such as nano- and microparticles, droplets, emulsions, and biological cells, are reviewed. The focus is on the in situ characterization of colloidal interfaces. Advantages of SFS/SHS, namely their interfacial selectivity and practical challenges, are also highlighted. This is not an exhaustive review but a short survey of the utility of SFS techniques through examples.

Let us first recap the interfacial specific knowledge we can gain from SFS and SHS methods. Table 7.1 summarizes the key information attainable. One major area of application is adsorption science – the process where molecules accumulate at an interface. The sum frequency intensity collected is proportional to the number density of molecules at an interface. This provides a direct connection to measure kinetics and isotherms in situ and in real time. SFS is rooted in molecular spectroscopy. Hence, interfacial electronic or vibrational spectra can be obtained. Polarization-dependent spectra not only help identify interfacial-bound chemical species but also monitor the relative changes in the spectral features, which also yield deeper insights into the local chemical and physical environment of the interface. Knowledge of orientation, chemical bonding and reaction, chirality, and randomness of the interfacial layer is acquired.

Given the sensitivity to interfacial potential and DC field-induced alignment of water molecules, SFS/SHS techniques can also elucidate charge properties of colloids. For colloids, scattering angle- and polarization-dependent SHS/SFS offer information pertinent to the alignment of water molecules around a charged interface. This can be further extended to study the ionization process, such as acid-base chemistry at an interface. Lasers used in SFG experiments are generally pulsed in the pico- to femtosecond regime. Thus, SFG pump-probe spectroscopy allows determining rates and molecular mechanisms of fast processes such as charge or energy transfers. Lastly, determination of $\chi_{ijk}^{(2)}$ components is also possible utilizing SFG, in particular, the HRS technique. The second-order susceptibility values are important in designing molecules, nanomaterials, and colloids of significance in optics and communication.

 | https://doi.org/10.1515/9783111018386-007

Table 7.1: Summary of application of SFG for interfacial studies.

Fundamental relationship	Possible information	Comments
– $I_{SFG} \propto N_s^2$ – $E_{SFG} \propto N_s$ – In situ and real-time measurements	– Adsorbate number density (N_s) – Thermodynamics of adsorption (ΔG_{ads}, ΔS_{ads}, ΔH_{ads}) – Adsorption kinetics ($N_s(t)$) – Molecular transport – Aggregation behavior	– Depends on the isotherm model applied; Determination of N_s may require information from other techniques – Signal interference and subtraction of signal from HRS and other mechanisms need to be considered – Temperature-dependent studies needed for ΔS_{ads} and ΔH_{ads} determination
– $I_{SFG} \propto \lvert\beta_{eff}^{(2)}\rvert^2$ – Frequency dependence of $\beta_{ijk}^{(2)}$	– Interfacial composition and molecular functional group identity – Chemical sensing of the interfacial environment (e.g., polarity) – Chemical interaction (e.g., H-bonding)	– Vibrational modes can be probed using VSFG – The extent and direction of spectral shifts relative to a reference system yield insights into the chemical makeup of the interfacial region or the type of bonds formed
– Relationship between $\chi_{ijk}^{(2)}$ and $\beta_{lmn}^{(2)}$ – Symmetry properties of $\chi_{ijk}^{(2)}$	– Molecular orientation and conformation – Interfacial chirality – Interfacial structure – symmetry breaking	– Involves measurement of SFG intensity at different polarization combinations. – In the case of SFS/SHS, polarization-dependent angle-resolved studies are necessary
– $I_{SFG} \propto \left\lvert \frac{\kappa}{\sqrt{\kappa^2 + (\Delta k_z)^2}} e^{i\varphi} \chi^{(3)} \psi_0 \right\rvert^2$ – Electrolyte-dependent studies	– Charge properties, e.g., surface potential and charge densities – Interfacial ionization process	– Phase measurement should be considered; heterodyne techniques needed. – In the case of SFS/SHS, angular dependence in conjunction with polarization studies are crucial
– $I_{HRS}(2\omega) \propto I^2(\omega)$ – $I_{SFG} \propto N_s^2$ combined with pump-probe spectroscopy	– Determination of second-order hyperpolarizabilities – Ultrafast dynamics – e.g., interfacial electron or charge transfer processes	– Require reference materials – Electric field-induced SHG can also be employed – Involves combining SFG/SHG experimental setup with the pump-probe technique (see Chapter 4)

7.1 Thermodynamics of Molecular Adsorption

Adsorption is the first step of any interfacial chemical process. The utility of SHG in selectively probing molecular adsorption in the backdrop of surrounding chemical species is exemplified in Figure 7.1(A). The strong SHS peak at 400 nm (fundamental wavelength 800 nm) is from an SHG dye adsorbed at the particle/aqueous interface. The dye and the particle solutions each produce weak HRS at 400 nm. By varying independent experimental variables (e.g., concentration and temperature), further insights into the adsorption process can be gained. To determine molecular adsorption affinity, an adsorption isotherm (see Section 3.6.1), which is a plot of adsorbate amount versus bulk equilibrium concentration of the molecule at a fixed temperature, is generated. Given that E_{SHG} or I_{SHG} is proportional to the number density of adsorbed molecules, N_{ads}, a plot of E_{SHG} or I_{SHG} versus the solution phase concentration of the adsorbate results in an adsorption isotherm. Polystyrene sulfate (PSS) microspheres and oil-in-water (O/W) emulsions were used as colloids for this pioneering work [118], which demonstrated the capability of SHG in determining adsorption energetics. Malachite green (MG), a triphenylmethane dye, served as the target compound. Figure 7.1(A) shows the chemical structure of MG and a schematic representation of its adsorption onto a PSS particle. MG has been commonly used for numerous SHG studies because the cationic form of MG has an electronic ($S_0 \to S_2$) transition around 425 nm in water. This happens to fall in the vicinity of the SH wavelength of a Ti:sapphire laser and thereby yields a resonantly enhanced SHG signal. The cationic dye has affinity for various negatively charged surfaces, such as the colloidal PSS particle. The SHG-based isotherm is reproduced in Figure 7.1(B). It shows a Langmuirian behavior with a sharp increase in SHG intensity at low dye concentration, representing molecular uptake, and then a plateau at a higher MG concentration as the interfacial sites become saturated.

Given the measurement is done in situ, the bulk concentration in this case is not the equilibrium concentration (c_{eq}) but that of the initial concentration (c) of MG. If depletion of the solution phase adsorptive concentration is negligible, i.e., $c_{eq}{\sim}c$, then the Langmuir model (see Table 3.1) can be applied directly. Since colloids provide a large overall surface area, combined with the fact that certain molecules can exhibit high affinity for the surface, the initial concentration can decrease considerably as adsorption occurs. This is the case for the isotherm collected for MG adsorption onto PSS particles. Thus, a generalized Langmuir model shown in eq. (7.1) has been applied to fit the data. In this equation, N_{max} is the maximum number density of the adsorbate, K is the equilibrium constant of adsorption, and c_0 is the concentration of the dispersant medium that serves as a reference point. For aqueous solution, $c_0 = 55.5$ M. Recall, $\theta = \frac{N}{N_{max}}$. Thus, the equilibrium constant is dimensionless, which allows calculation of the Gibbs free energy of adsorption as follows: $\Delta G_{ads} = -RT \ln K$. Compared to the classical Langmuir fit, the generalized model shows a better representation of the experimental data, as seen in Figure 7.1(B):

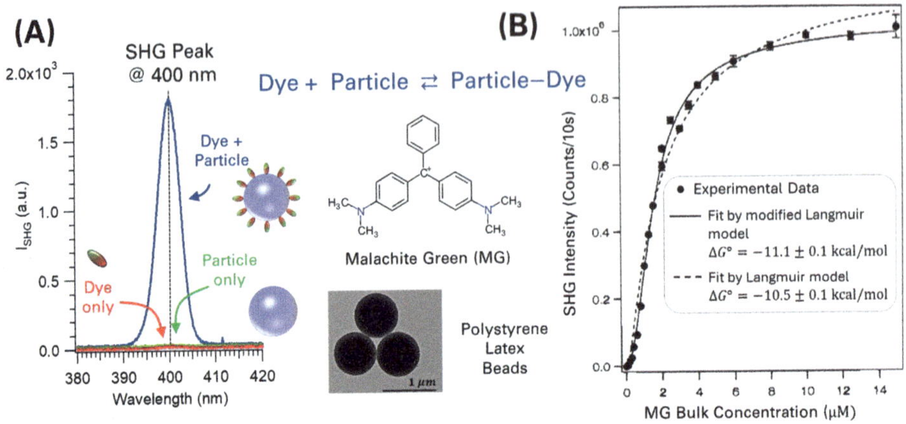

Figure 7.1: (A) SHG signal from particle-dye mixture compared to signal from dye and particle by itself. Also shown is the structure of malachite green (MG and TEM image of polystyrene sulfate (PSS) latex particles. (B) SHG-based adsorption isotherm [118] of MG in the presence of PSS particles. Adapted and reprinted with permission from Wang et al. [118]. Copyright 1998, American Chemical Society.

$$\theta = \frac{\left(c + N_{\max} + \frac{c_0}{K}\right) - \sqrt{\left(c + N_{\max} + \frac{c_0}{K}\right)^2 - 4cN_{\max}}}{2N_{\max}} \tag{7.1}$$

With the use of the generalized Langmuir model, it is possible to obtain the maximum surface density of the adsorbate. The $N_{\max}$ parameter does not appear in the classical Langmuir model. Caution is warranted in overfitting the data using eq. (7.1) as it applies only to the situation where bulk concentration depletion is significant. It does not apply to the case when the bulk concentration is much greater than the adsorbate concentration (c_{ads}). If $c_0 \gg c_{\text{ads}}$, eq. (7.1) does not yield accurate values of $N_{\max}$. It is plausible that a colloidal particle surface is composed of multiple chemically distinct surface sites. In such a scenario, the generalized multicomponent Langmuir (Table 3.1) model is applicable, and this has been demonstrated using SHG to probe multisite adsorption [119]. The effect of the surface functional group on the adsorption of MG onto polystyrene microspheres at various pH levels has also been investigated [120] using SHS.

The adsorption studies discussed so far involved particles that do not themselves generate a coherent SHG signal. Hyper-Rayleigh scattering at 2ω from these colloids is generally weak and is subtracted from the resonant SHG signal originating from the surface-bound species. There are situations where the bulk or the neat surface of the colloidal substance is intrinsically anisotropic. Examples of such colloids include metallic particles, crystalline mineral oxides, and, as highlighted in this section, carbon black. Coherent SHG is generated from the bulk and the surface of the particles, which in turn can interfere with the SHG signal originating upon molecular adsorption.

To analyze adsorption processes in this case, the following equation describes the overall SHS considered:

$$I_{SHS} = B + \left| b + a\theta e^{i\varphi} \right|^2 \tag{7.2}$$

In this equation, B represents all possible incoherent contributions (e.g., HRS, residual fluorescence, etc.) to the detected signal. This background signal is usually negligible compared to the coherent SHG. In practice, this background can be subtracted based on SHS measurements from control samples, such as the dye or the particle solution itself. The terms $a\theta e^{i\varphi}$ in eq. (7.2) corresponds to the contribution from the adsorbate, whereas b is the coherent contribution from the bulk or clean surface. The surface contribution scales with fraction adsorbed, θ. The factor $e^{i\varphi}$ represents the phase difference between the SHG signals originating from the particle versus the adsorbate. The signals can either add or subtract. We see a glimpse of such a phenomenon in Figure 7.2(A). It shows an SHG-based adsorption isotherm of MG onto carbon black nanoparticles (c.a. 160 nm) of different oxygen/carbon (O/C) ratios. For instance, CB 1 corresponds to an O/C ratio of 0.028, whereas CB14 has a ratio of 0.270. The colloidal mixture is in an aqueous suspension of pH 4 to ensure the cationic form of MG. Each curve has been normalized by dividing the SHG intensities at each MG concentration by that of zero concentration. These traces show the effect of interference. In certain cases, the normalized SHG intensity dips below 1, an effect of destructive interference between the signals from the two different sources. It is also apparent that the SHG isotherm profiles are different from each other. This is the effect of distinct surface chemistry (different O/C ratio). This highlights the surface sensitivity of SHG.

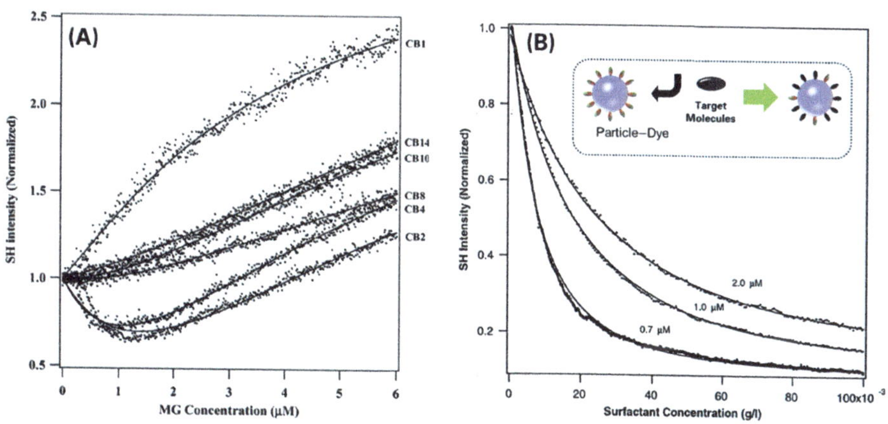

Figure 7.2: (A) MG adsorption onto carbon black particles of different O/C ratios as probed using SHG. Adapted and reprinted with permission from Wang et al. [121]. Copyright 2007, American Chemical Society. (B) Surfactant adsorption isotherm [122] onto PSS. Adapted and reprinted with permission from Wang et al. [118]. Copyright 2000, American Chemical Society.

These examples of adsorption isotherms involve molecules that are resonantly matched with the second harmonic frequency. It is not common to find molecules with overlaps between their electronic transition frequency and the laser or the SHG frequencies. Lasers are often limited to a certain narrow wavelength range. One approach, as in the case of electronic sum frequency generation (ESFG), is to use broadband lasers with a larger tunability range. ESFG allows the collection of interfacial spectra in one shot and has been applied to study molecular adsorption at planar interfaces [123, 124]. A challenge, however, is that the ESFG experimental setup is complex and requires spatial and temporal overlapping of multiple laser beams. Moreover, even if there exists a resonant overlap, molecules may not have a large second-order susceptibility and thereby not yield a strong SHG or SFG signal. General requirements for a large molecular $\chi_{ijk}^{(2)}$ are a large transition dipole moment and a large difference between the excited- and ground-state dipole moments. Pi-conjugated systems, with strong electron-donating and electron-accepting groups, having a push-pull architecture, typically satisfy these criteria. Examples of common organic molecules that exhibit nonlinear optical properties and have been utilized for SHG studies are summarized in Table 7.2.

Not all molecules have highly polarizable chromophores. In this case, adsorption of such molecules can be studied based on competitive adsorption with the aid of a detectable SHG probe molecule. The basic idea is that a SHG-active dye (Table 7.2) is used to saturate the particle interface. This saturated coverage results in a constant SHG signal. Then a nonresonant target molecule, which would not generate SHG on its own, is introduced into the colloidal suspension. If the target molecule has a higher affinity for the colloidal interface, it displaces the probe molecule from the surface. The SHG signal decreases as the concentration of the target compound is increased. An example of this type of study is depicted in Figure 7.2(B). The inset is a cartoon depicting the displacement of the dye molecule by the target compound. In this work [122], surfactant adsorption on colloidal latex particles has been studied with the aid of a SHG probe molecule, MG. Surfactants do not generally have chromophores that are SHG active. The ΔG_{ads} and surface density of the probe molecule is determined first, as described previously. These parameters, along with the extent of the decrease in the SHG signal, are then used to determine ΔG_{ads} and the surface density of the surfactants. Assuming a Langmuirian behavior, the multicomponent adsorption isotherm model (Table 3.1) is applied. A similar strategy was applied [125] to study the adsorption of amino acids, proteins, and other macromolecules.

A clear advantage of SHS is that adsorption isotherms can be measured in situ, without the need to separate the particles from the solution. Small colloids and particles of low density are difficult to separate. A few more SHS adsorption studies are presented in this section to highlight the versatility of this technique. Emulsions are liquid-in-liquid colloids and are not possible to study based on separation techniques. Early studies [95, 118, 126] have shown the utility of SHS for colloidal O/W emulsions. An example of dye adsorption onto hexadecane O/W nanoemulsions (NEs) of radius

Table 7.2: List of common SHG dye molecules.

Compound class	Examples	Comments
Triphenyl-methane dyes	– Malachite green (MG), crystal violet (CV) 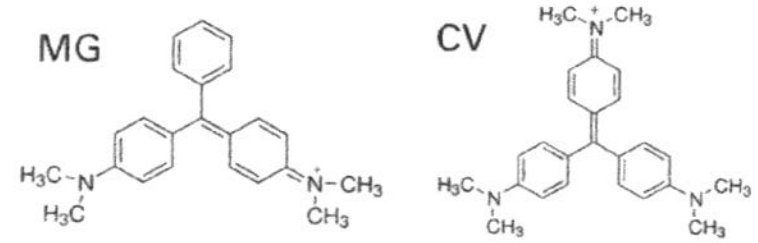 	– The cationic form is pH dependent – The first SHG experiment for molecular adsorption on colloids was performed using MG
Styryl dyes	– 4-(2-[6-(Dioctylamino)-2-naphthalenyl]ethenyl)-1-(3-sulfopropyl)pyridinium salt (Di-8-ANEPPS) – *trans*-4-[4-(Dimethylamino)styryl]-1-methylpyridinium iodide (DASPI, DSMI) 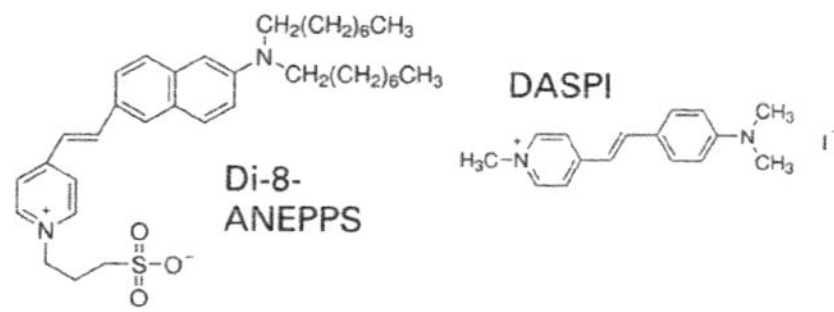 	– Different alkyl chain lengths and configurations can be used to study dye-membrane interactions – Has been used with colloids – Also applicable for SHG microscopy
Coumarin dyes	– Coumarin 314, Coumarin 343 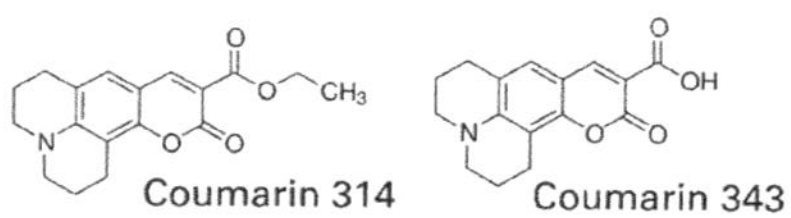 	– Coumarin dyes can be sensitive to pH or solvent polarity

(continued)

Table 7.2 (continued)

Compound class	Examples	Comments
Stilbazolium salts	– 4′-Dimethylamino-*N*-methyl-4-stilbazoliumtosylate (DAST) – 4 *N*, *N*-Dimethylamino-4’-*N*’-methylstilbazolium 2,4,6-trimethylbenzenesulfonate (DSTMS) 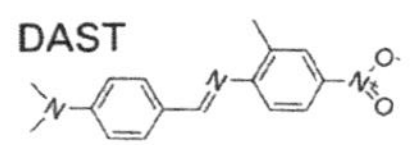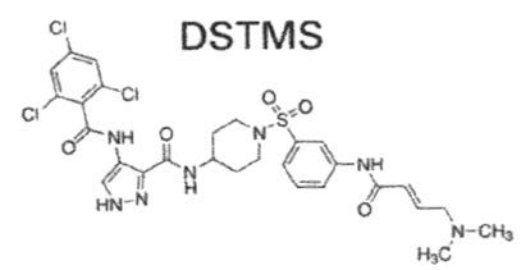	– Used for NLO crystals – Performs well in longer IR wavelength regions
Miscellaneous dyes	– Methyl orange – Naphthol yellow S (NYS) 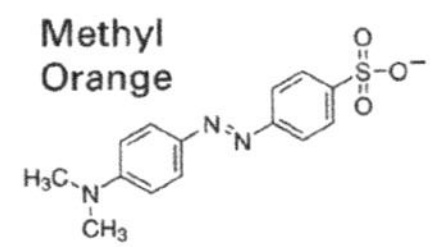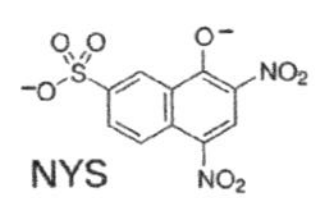	– NYS and methyl orange are anionic – Additional dyes include rhodamine 6G, fluorescein derivatives, and oxazolyl dyes.

115 nm, ca., is shown in Figure 7.3(A). In this study, NEs with cationic, anionic, neutral, and zwitterionic surfactants were synthesized. Adsorption of positive, negative, and neutral organic molecules to these NEs was investigated. It was found that Coulombic attraction plays a key role, but the presence of a charged interface does not necessitate molecular adsorption at these NE oil/surfactant/aqueous interfaces. This work demonstrates that molecular adsorption based on Coulombic interaction can and does vary depending on the charge structure of the interface, i.e., the source and the type of charge accumulation.

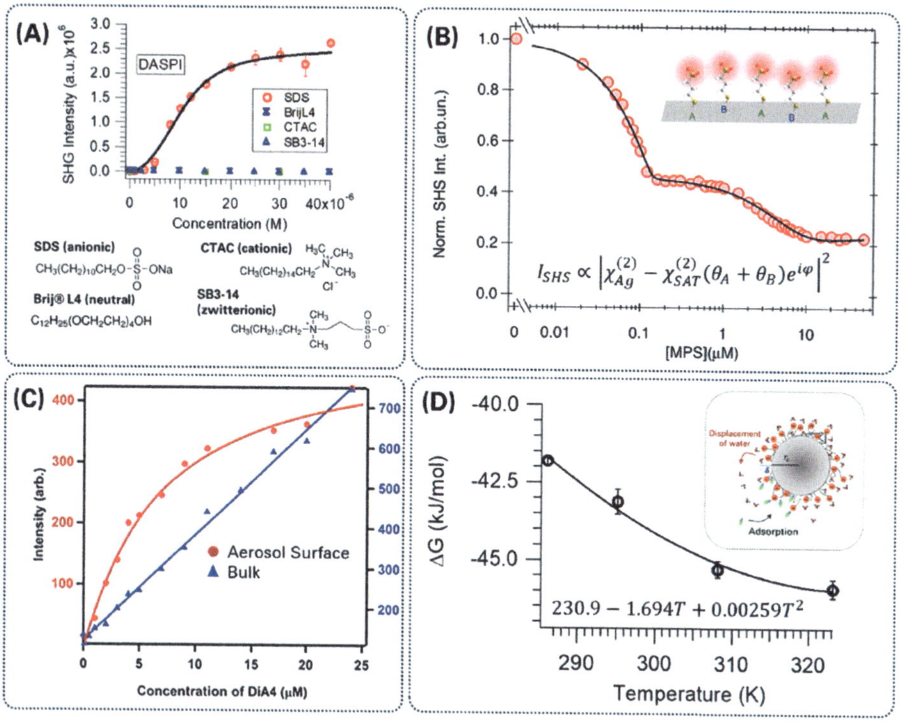

Figure 7.3: (A) Adsorption of DASPI, a cationic styryl dye, onto hexadecane O/W nanoemulsion composed of different surfactants. Adapted and reprinted with permission from Sara et al. [127]. Copyright 2024, American Chemical Society. (B) Normalized SHG adsorption isotherm of 3-mercaptopropanesulfonate (MPS) onto citrate-capped Ag nanoparticles. Adapted and reprinted with permission from Xu et al. [128]. Copyright 2015, American Chemical Society. (C) Coherent SHS intensity (red circle) and incoherent HRS of bulk liquids (blue triangle) as a function of DiA4 concentration. The red curve is a model fit for the SHS signal using the Langmuir isotherm model, whereas the blue line represents a linear behavior expected for the bulk-originated HRS intensity dependence on concentration. Adapted and reprinted with permission from Wu et al. [129]. Copyright 2016, American Chemical Society. (D) Adsorption free energy versus temperature as determined by SHS for molecular DASPI adsorption onto polystyrene particle. Adapted and reprinted with permission from Kanu et al. [130]. Copyright 2025, American Chemical Society.

Adsorption of thiol compounds onto metallic Ag nanoparticles (diam. 80 ± 20 nm) has been explored [128] using SHS. Like adsorption on carbon black (see Figure 7.2 (A)), the neat silver nanoparticles yield an SHG signal. In this case, upon adsorption of the thiol compound, MPS, the SHG signal decreases. A representative normalized isotherm is shown in Figure 7.3(B). Since there are two sources of SHG, eq. (7.2) is applicable. The parameters a and b are $\chi^{(2)}_{Ag}$ and $\chi^{(2)}_{SAT}$, respectively, where $\chi^{(2)}_{Ag}$ corresponds to the susceptibility of the bare Ag nanoparticle and $\chi^{(2)}_{SAT}$ is the Ag surface susceptibility induced by MPS chemisorption. It is clear from Figure 7.3(B) that the adsorption of MPS on the nanoparticle/aqueous interface proceeds via two-step mechanism, primarily a result of the charge-charge repulsion among the anionic MPS adsorbed on or near the surface. In addition to the isotherm, this insight was gained based on temperature-dependent kinetic studies (as discussed in Section 7.2) of MPS adsorption onto these particles. This study highlights the significance of repulsive interactions, both electrostatic and steric in nature, in the chemisorption of charged thiols on metallic nanoparticles. SHS, in conjunction with complementary spectroscopic techniques, has been further applied to investigate metallic colloids. These studies [131, 132] include the investigation of surface interactions, including biomolecules, and understanding plasmonic and charge effects of nanoparticles.

Application of SHS has been pushed even further in recent years. Figure 7.3(C) shows an example isotherm (red markers) of organic dye (DiA4) adsorption onto laboratory-generated aerosol particles suspended in air. This work [129] demonstrates a new approach for characterizing aerosol particle surfaces using SHG techniques. An adsorption isotherm at a fixed temperature allows for the determination of ΔG_{ads}, the molecular affinity toward an interface. Isotherms can be generated at different temperatures. For instance, ΔG_{ads} of the DASPI onto polystyrene microspheres (representative of model microplastics) have been determined [130] at various temperatures. An example data, which shows a nonlinear dependence of ΔG_{ads} with temperature, is reproduced in Figure 7.3(D). Thermodynamics dictates $\Delta S_{ads} = -\left(\frac{\partial \Delta G_{ads}}{\partial T}\right)_P$. Thus, the negative slope reveals that there is an increase in entropy upon molecular adsorption at the aqueous interface of colloidal microplastics, a telltale sign that adsorption is associated with the displacement of interfacial bound water molecules. This work shows that the interfacial distribution of water and surface functional group-dependent interactions determine the magnitude of binding strengths.

7.2 Adsorption Kinetics and Molecular Transport

With fast signal detection and in situ capability, SHS is a valuable tool for studying real-time kinetics of adsorption involving colloids. It allows investigation of not just the adsorption process but also interfacial chemical reactions and molecular transport. Fundamental insights, such as rate constants, activation energy, and mechanisms of adsorption, can be obtained from kinetic measurements. As an example, the adsorption kinetics of thiol compounds on Ag nanoparticles at different temperatures have been determined [128]. Figure 7.4(A) shows the kinetic profiles (top panel) and the variation of the natural log of the rate constant, $\ln k$, with inverse Kelvin temperature (bottom panel). According to the linear form of the Arrhenius equation, $\ln k = \ln A - \frac{E_a}{RT}$. The slope of this plot is thus used to calculate the activation energy (E_a) of the adsorption. This study showed that there are two distinct activation energies as thiol adsorption occurs via two different mechanisms. At low surface coverage (phase A), where intermolecular repulsion is negligible, the E_a is lower compared to that of high surface coverage (phase B).

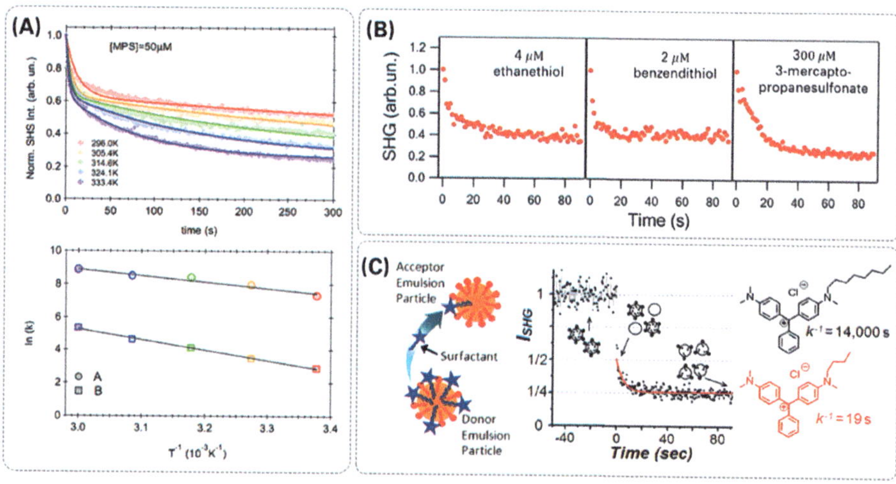

Figure 7.4: (A) Temperature-dependent adsorption kinetics of MPS onto Ag nanoparticles. Normalized SHG versus time (top panel) and $\ln k$ versus T^{-1} (bottom panel). Rate constants associated with adsorption at sites A and B are represented as circles and squares, respectively. Adapted and reprinted with permission from Xu et al. [128]. Copyright 2015, American Chemical Society. (B) Different extent of decrease in SHS intensities from Ag nanoparticle due to addition of three different thiol compounds. Adapted and reprinted with permission from Gan et al. [133]. Copyright 2011, American Chemical Society. (C) Transfer of surfactant molecules within emulsion particles as probed by SHS. Adapted and reprinted with permission from You et al. [134]. Copyright 2012, American Chemical Society.

Recall, the source of SHS from metallic nanoparticles can include the neat particle and the surface adsorbate. However, using Ag nanoparticles, it has been shown [133] that SH intensity substantially decreases upon chemical bonding of thiol molecules. Growth dynamics of colloidal gold-silver core-shell nanoparticles have also been observed using SHS [135]. Figure 7.4(B) highlights the chemisorption of three different thiol compounds onto the Ag nanoparticle and thereby the surface sensitivity of the signal. It has been suggested that most of the SH intensity detected from the Ag nanoparticles is generated from the surface layer of Ag atoms. SHG to investigate the adsorption of molecules possessing small nonlinear polarizability at the surface of metallic nanoparticles is thus possible. SHS investigation of ligand exchange at colloidal gold demonstrated two reaction processes. Fast and slow processes were attributed to the adsorption of a thiol compound at low- and high- coordinated Au nanoparticle surfaces. SHG kinetic experiments have also been performed to monitor the transfer of surfactant molecules in emulsions. In this clever study [134], emulsions impregnated with different SHG probe surfactants were synthesized. Use of different types of probe molecules revealed that an increase in the hydrophobic chain length decreases the transfer rate.

Phenomenologically, many SHS studies have found biexponential functions to be suitable in fitting adsorption kinetic data. Adsorption onto two different sites or two separate processes can explain such phenomena. However, it is important to recognize that adsorption onto colloidal interfaces is complex. There could be multiple energetically distinct sites. Adsorption into an interfacial layer can be diffusion-limited. Interaction between adsorbates can occur. These factors may render existing kinetic models (as sampled in Table 3.2) not appropriate to describe SHG-based kinetic data. It is also worth emphasizing that there could be different sources of coherent SHG from a colloidal suspension. For example, both the bulk and the surface (as in carbon black or metallic colloids) can contribute. These SHG fields are likely to interfere, and eq. (7.2), which includes the phase difference, is applicable. It is also possible that molecules adsorb on different surface sites with different orientations. Interference in signals originating from each site can lead to SHG cancellation.

Consider the case of adsorption at two different sites, both described by Langmuirian kinetics (Table 3.2). The overall evolution of SHS signals with time, normalized by the SHS signal at $t = \infty$, can be written as follows:

$$\frac{I_{\mathrm{SHS}}(t)}{I_{\mathrm{SHS}}(t=\infty)} = \left|\theta_{\max,1}\left(1-e^{-(k_1 c_0)t}\right) + \theta_{\max,2}\left(1-e^{-(k_2 c_0)t}\right)e^{i\varphi}\right|^2 \tag{7.3}$$

The maximum fraction of molecules at site 1 and site 2 is described as $\theta_{\max,1}$ and $\theta_{\max,2}$, respectively. Note, $\theta_{\max,1} + \theta_{\max,2} = 1$. As in eq. (7.2), $e^{i\varphi}$ is the phase factor describing the interference. Figure 7.5(A) shows the effect of different surface coverage for a fixed φ (left panel) and the effect of φ for a fixed $\theta_{\max,i}$ (right panel). When $\varphi = 180$, the signals originating from two dipoles are completely out of phase. If adsorption on site one is faster than that of the second site, signals will rise to a maximum and then decrease as the second site populates. One way this scenario is possible is if the dipoles of the two adsorbates are oppositely oriented. This concept is precisely useful in applying SHS to study transmembrane crossing across a colloidal lipid bilayer. It was first demonstrated with liposomes, a spherical vesicle made of phospholipid bilayers with an enclosed aqueous core. Colloidal liposomes serve as a model for biological cells. Figure 7.5(B) shows a cartoon in which the transport properties of a small molecule across a liposome membrane are depicted. This original work [136] used MG as the SHG probe to demonstrate its transport into liposomes. Subsequently, the effect of various biological molecules, such as cholesterol, peptides, and antibiotics, on the membrane transport of MG and ions has been studied using SHS [137, 138].

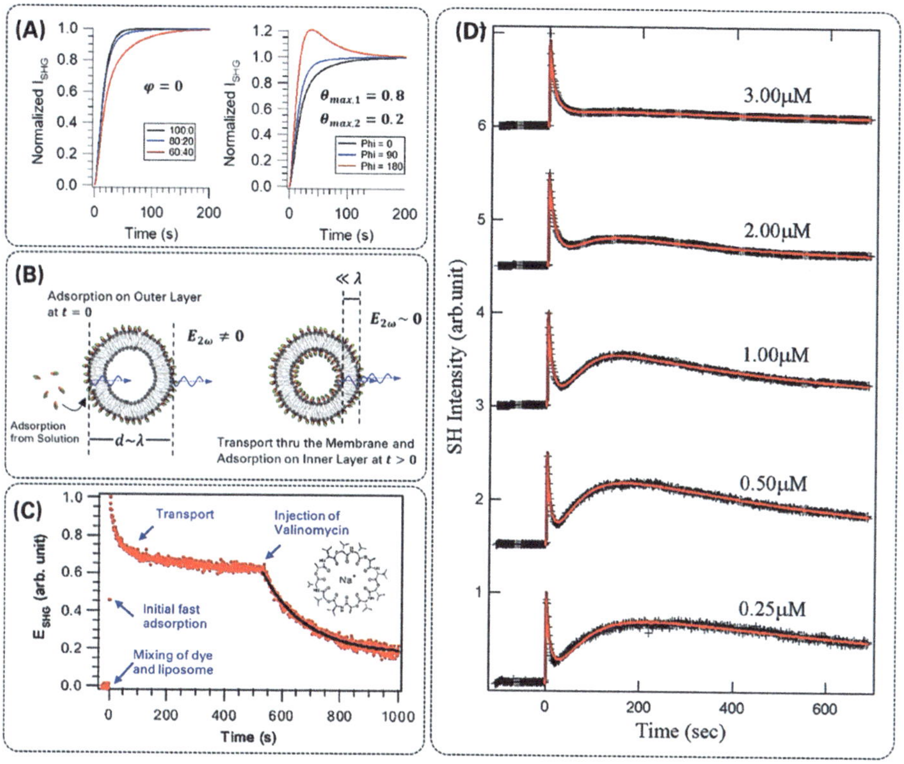

Figure 7.5: (A) Simulation of eq. (7.3), highlighting the effect of surface coverage with constructive interference (left panel) and various interference at a fixed surface coverage (right panel). (B) Transport of molecules across a colloidal liposome or biological lipid membrane and the effect on SHG from molecules adsorbed at outer and inner leaflets. (C) Ion transport across the liposome bilayer, and the effect of valinomycin, as probed using SHS. Adapted and reprinted with permission from Liu et al. [138]. Copyright 2008, American Chemical Society. (D) MG crossing *E. coli* cell as probed by SHG signal. Adapted and reprinted with permission from Zeng et al. [139]. Copyright 2008, American Chemical Society.

Figure 7.5(C) illustrates an example data of transport kinetics probed using SHS. Transport of MG into dioleoyl phosphatidylglycerol liposomes suspended in 10 mM NaCl solution is shown. At $t = 0$, as cationic MG is mixed with liposome, MG adsorbs very rapidly onto the external leaflet of the liposome, leading to a sharp initial rise in SHG. Immediately after, SHG intensity is observed to decrease with time. This is due to the transport of MG into the interior of the vesicles and its adsorption onto the inner layer of the membrane. MG adsorbed at the inner leaflet is oppositely oriented to that of the MG adsorbed at the outer layer. Given that the thickness of the lipid bilayer is in the order of a few nanometers, the signals originating from these two interfaces are out of phase. The overall signal decreases as the inner leaflet populates. After an equilibrium is reached, valinomycin (Val), a cyclic peptide antibiotic that functions as a potassium ionophore, is added, and a further decay in signal is observed. It appears that more cationic MG enters the liposome and populates the inner layer. The driving force of additional transport of MG has been attributed to the transport of Val-mediated Na^+ ions out of the liposome. The study of molecular and inorganic ion transport using SHS is not limited to model systems anymore. Recent years have demonstrated that SHS can be applied to real, living biological cells not only to study transport but also changes in the membrane structures [139–144].

With more complex systems, elaborate SHS kinetic profiles may be observed. Figure 7.5(D) is an example, which shows time-profiles of normalized SH intensity after the addition of MG at various concentrations into solutions containing living *Escherichia coli* cells [139]. It is apparent that not just adsorption onto the outer and inner leaflets is occurring, but the kinetics involve multiple processes. An *E. coli* bacterium is about 1–2 μm and contains an outer membrane and an inner (cytoplasmic) membrane. Typical thickness of each layer is approximately 6–8 nm, and the space between these two bilayers, periplasms, can range from 10 to 25 nm. Molecules, such as MG, can bind to a total of four different interfaces. Given the discussion on phase difference, various forms of constructive and destructive interference are plausible. Figure 7.6(A) illustrates how the SHG intensity can vary depending on the transport mechanisms involved.

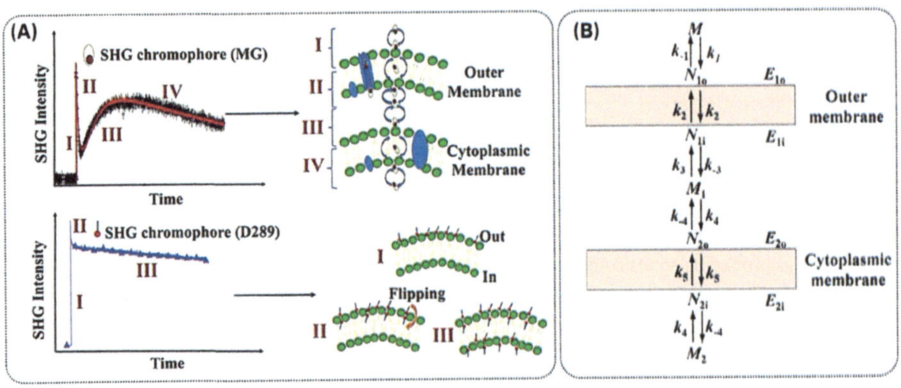

Figure 7.6: (A) Typical SHS kinetic profiles observed and plausible correlation considering adsorption, transport, and orientational changes that may occur. (B) A kinetic model depicting adsorption, desorption, and transport processes of an arbitrary concentration of an organic molecule (M) across outer and cytoplasmic membranes of *E. coli*. The values $k_{\pm1}$, $k_{\pm2}$, $k_{\pm3}$, and $k_{\pm4}$ represent adsorption and desorption rate constants, whereas k_2 and k_5 are transport rate constants. The symbols E and N correspond to empty site and number of molecules at these inner (i) and outer (o) layers. Adapted and reprinted with permission from Chen et al. [144]. Copyright 2024, Elsevier.

The SHG kinetics observed depend on the SHG probe molecule used and the type of vesicles involved. For instance, the top panel shows MG transport across *E. coli*, whereas the bottom panel is an example of D289, a styryl dye, interaction and transport across dioleoyl phosphatidylglycerol liposomes. For the latter scenario, these studies suggest that orientational flipping could explain the observed kinetics. Adsorption and transport kinetic models are often used to fit kinetic data. An example model applicable to *E. coli* is shown in Figure 7.6(B). A set of four differential equations can be written based on this model to describe the change in SHG molecule densities (N_{1o}, N_{1i}, N_{2o}, and N_{2i}) at the four distinct interfaces. In accordance with eq. (6.8), the total SHG intensity from all the surfaces can be described as $I_{SHG} \propto (N_{1o} - N_{1i} + N_{2o} - N_{2i})^2$. It is assumed that the outer and inner populations of the dye are completely out of phase. This model has been used to fit the SHG curves (Figure 7.5(B)) and obtain the maximum adsorption density of molecules on the interfaces and the associated rates and the free energies of the adsorption, which are determined by the k values.

SHS has been applied to study adsorption processes onto various other colloidal substances [95, 145]. In recent years, the implementation of this technique in areas, including but not limited to biomedical, nanomaterial, atmospheric, and environmental sciences, has emerged. The examples highlighted herein capture the essence of the type of adsorption measurements and fundamental insights that can be achieved using this light scattering technique. Adsorption isotherms and kinetics are usually obtained at a fixed wavelength and provide information about the chemical processes. SFS interfacial spectra, which we turn to next, provide structural information about colloidal interfaces.

7.3 Colloidal Interfacial Spectroscopy

Electronic transitions in molecules occur in the ultraviolet and visible range of the spectrum. ESFG and SHG are utilized to probe these transitions in molecules adsorbed at an interface. VSFG is used to probe vibrational modes in the IR region. Collection of SHG and VSFG spectra from various planar interfaces is common, but thus far has been limited in the case of colloids. Scattering from colloids requires complex mathematical models. Complex size-dependent light scattering, background signal from the bulk medium, and frequency-dependent attenuation of scattered light make it experimentally challenging to collect spectra from a colloidal suspension. Yet, spectral features provide molecular identity and a wealth of information about interfacial structure. While these challenges have curbed the investigation of colloidal systems, ongoing fundamental research (see Section 6.3) pushes the limit in using specialized techniques and engineered colloids to advance our knowledge of colloidal interfaces. Application of vibrational sum frequency scattering (VSFS) for spectral determination is on the rise. In this section, representative spectra collected from colloidal interfaces are presented. The intention is to highlight the key information that can be achieved from spectral interpretation.

7.3.1 Electronic Spectroscopy at Colloidal Interfaces

The charge-transfer band of a surface complex formed on TiO_2 colloidal particles has been observed using SHS. In this experiment, a tunable femtosecond Ti:sapphire laser was used. Laser wavelength was tuned in the range of 750 to 920 nm, and the corresponding SH signal was detected. The signal was corrected for turbidity and other attenuations along the light path. The SH spectrum (Figure 7.7(A)) of catechol adsorbed on anatase TiO_2 (~400 nm) colloidal particles in aqueous suspension revealed a charge-transfer band centered at 2.72 eV (456 nm). Surface photosensitization of TiO_2 is a very important process in solar energy conversion, and this study allowed direct observation of the interfacial charge-transfer complex.

Likewise, the interfacial spectrum ($S_0 \rightarrow S_2$ transition) of MG adsorbed at carboxylate polystyrene/aqueous interface was recorded. The effect of photolysis using UV-visible light was studied. It was demonstrated that the photodegradation proceeds via an *N*-demethylation mechanism at the colloidal/aqueous interface based on a newly generated peak that manifests as a shoulder slightly blue-shifted relative to the parent compound (Figure 7.7(B)). The difference between the photolyzed and unexposed MG spectra further revealed the identity of the intermediate species to be monodes-MG, one of the demethylated intermediate species. Figure 7.7(C) shows an MG spectrum at the surface of aerosol particles. In this case, the MG spectrum ($S_0 \rightarrow S_1$ transition) at the interface is blue shifted. These studies highlight the significance of SHS/ESFS in identifying chemical species at colloidal interfaces. In situ SHS and HRS measurements from different types of nanoparticles in the aerosol phase have also been demonstrated [148]. The potential for measuring the specific surface area of an aerosol using SHS as laser-based diagnostics has been proposed.

SHS experiments can shed light on the polarity of the interfacial environment. Certain molecules are solvatochromic, i.e., sensitive to the polarity of the solvent. If such a molecule is adsorbed at an interface, collecting its spectrum and comparing it with those measured in other polar and nonpolar media can provide the relative polarity of the chemical interface. Consider a case when the dipole moment of the molecule in its ground state is larger than that of the excited state ($\mu_{gs} > \mu_{es}$), as shown in Figure 7.8(A). The larger the dipole, the better it is stabilized by a more polar solvent. The transition involves a larger energy gap. As the polarity of the solvent is increased, the electronic transition occurs at a shorter wavelength. The spectrum exhibits a blue-shift or hypsochromic shift, as shown in Figure 7.8(C). This is called negative solvatochromism. Molecules exhibiting proton transfer chiefly display negative solvatochromic behavior. In contrast, positive solvatochromism occurs when the dipole moment of the molecule in its excited state is larger than that

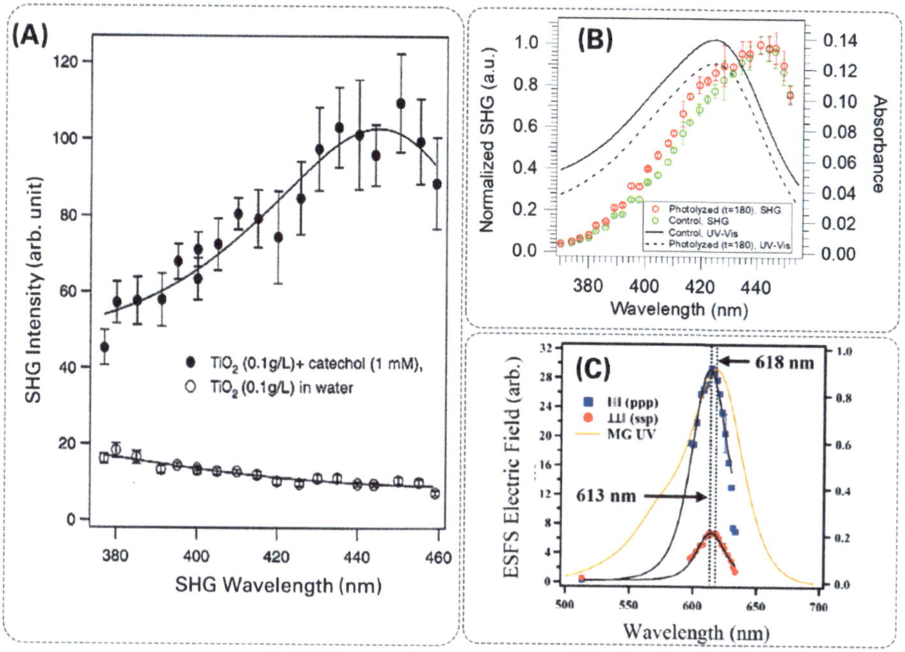

Figure 7.7: (A) Normalized SH spectra of TiO_2 in water, TiO_2 in 1 mM catechol aqueous solution, and TiO_2 in water. Adapted and reprinted with permission from Liu et al. [146]. Copyright 1999, American Chemical Society. (B) Surface SHG spectra (markers, left axis) of MG before (green) and after photolysis (red) overlaid with UV-Vis absorbance spectra (traces, right axis) of MG at the photolysis exposure time of $t = 0$ (solid) and $t = 180$ (dashed). The x-axis represents the UV-Vis and the SHG wavelength for the bulk and surface spectra, respectively. Adapted and reprinted with permission from Kaylor et al. [147]. Copyright 2022, Elsevier. (C) ESFS spectra of MG from NaCl aerosol particles as compared with UV-Vis absorption of MG in bulk solution (solid line). The solid lines (black) are fitted results by using eq. (S1). SSP, red solid circles; PPP, blue solid squares. Adapted and reprinted with permission from Qian et al. [102]. Copyright 2020, American Chemical Society.

of the ground state ($\mu_{es} > \mu_{gs}$) during an electronic transition. The energy of the excited state is decreased, and the transition energy is lowered (Figure 7.8(B)). This results in a red shift in the spectrum (Figure 7.8(C)), also known as a bathochromic shift. Typically, this is the case for π to π^* transitions. Charge-transfer active molecules often exhibit positive solvatochromism.

The concept of a polarity scale using solvatochromic dyes is well known [149, 150]. It has been applied in conjunction with SHG spectroscopy to infer the polarity of planar interfaces such as air/water and air/oil interfaces [151]. The MG spectra shown in the top panel of Figure 7.7(B) demonstrate this for the colloidal interface. The interfacial spectrum (green markers) is red-shifted compared to its spectrum in water (solid black curve). In previous studies, the bathochromic shift in MG has been correlated with a more polar environment. Accordingly, the red shift observed suggests an overall polar

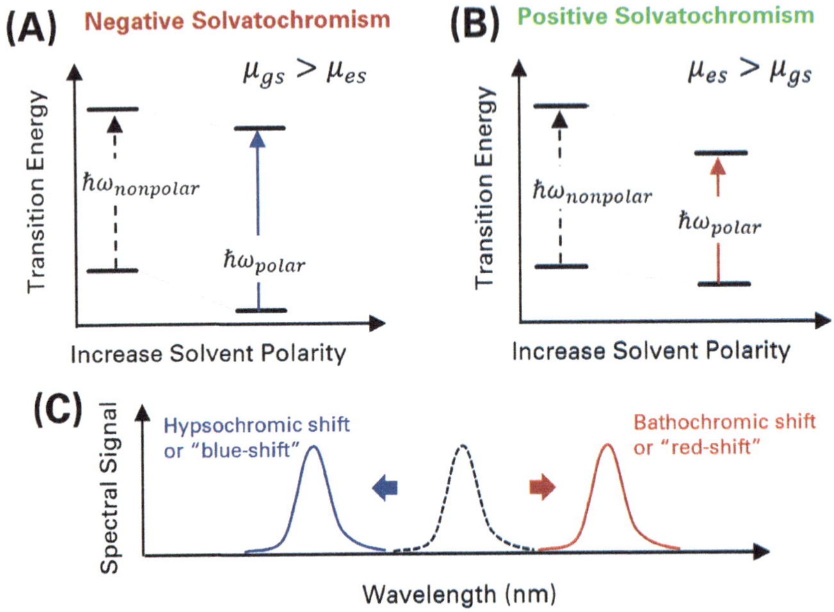

Figure 7.8: Illustration of the concept of solvatochromism and spectral shift.

environment at the negatively charged PSC/aqueous Interface [147]. A hypsochromic spectral shift of MG adsorbed at the aerosol surface (Figure 7.7(C)) reveals a nonpolar aerosol/air interface. Thus, one can conceive the use of probe molecules that are sensitive to not just polarity but also other external stimuli such as pH, temperature, and voltage. This creates the potential for chemical sensing of various colloidal interfacial environments using SHS.

However, it has been reported that structurally different molecules "sensed" different polarities at the air/water interface [152]. ESFG spectra of five different coumarin dyes showed that the local solvation environments around different coumarin molecules are different at the air/water interface. As a result, the effective polarity could not be defined as an intrinsic property of the air/water interface. A small difference in the orientation angle and vertical position of the solute at the water interface led to a substantial difference in the stabilization energy and thereby spectral shift. The spectrum is highly sensitive to the local environment. This study highlights that the interpretation of SFG spectra is complex but can be informative about interfacial structure.

7.3.2 Vibrational Spectroscopy at Colloidal Interfaces

The examples highlighted within the context of SHS or ESFS involve the use of resonant dyes as the probe molecule. VSFS spectroscopy, on the other hand, can directly "see" the chemical functional groups that are at or in the vicinity of the colloidal interface. This is based on probing the vibrational modes of known chemical functional groups. Thus, VSFS serves as a label-free approach to gain insight into the interfacial chemical structure. A common approach in studying particles using VSFG has been to deposit them on a flat substrate and perform experiments in reflection geometry. Such investigations [153–157] have been conducted on surfactant-based particles, metallic nanomaterials, sea spray aerosols, and organic particulate matter on indoor surfaces, as well as inorganic minerals. While these investigations provide critical information on particle surface composition and molecular orientation, they do not reflect interfacial behavior when particles are suspended. Herein, VSFS studies involving colloidal suspensions are presented. Interfacial insights are gained from particles in their colloidal state.

Representative data that demonstrated [158] VSFS from colloidal suspension for the first time is shown in Figure 7.9(A). The colloidal system in this experiment is ~340 nm SiO_2 particles coated with stearic alcohol ($C_{18}H_{37}OH$) dispersed in CCl_4. The VSFG experimental setup involves IR and visible pulses centered around 2,900 cm^{-1} and 800 nm, respectively. The solvent CCl_4 is centrosymmetric and transparent in the IR frequency range of interest (see Figure 4.3). It does not produce its own SFG signal in the spectral window of interest and allows the SFG signal from the particle surface to pass through the suspension and be detected. SiO_2 does not absorb IR light in this range as well. Figure 7.9(A) shows vibrational spectra corresponding to CH stretches (Figure 4.3) at four different polarization combinations. Equation (6.11) is used to fit the spectral profile. Figure 7.9(B) is the VSFS scattering pattern from the particles. Based on RGD approximation, the authors obtained information about the local molecular order and density of the alkyl chains adsorbed at the SiO_2 particle surface. More recently, VSFS has been applied to study interfacial spectra of colloidal metal-organic frameworks (MOFs) [159]. Figure 7.9(C) provides an example VSFS spectrum of Zn(2-methylimidazolate)$_2$ (ZIF-8) MOF in the C–H stretching region. These spectra revealed the presence of solvation shells around this MOF in aqueous, nonaqueous (DMF), and in solvent mixtures. Peak designations are provided in the illustration of the linker vibrational modes. In polar solvents, the presence of solvent-metal binding and spontaneous ordering of organic bridging linkers within the MOF has been observed. Since MOFs are porous in nature, they challenge conventional models of colloidal and interfacial chemistry. The results of this study help explain the unexpected colloidal stability of MOF.

VSFS is an emerging tool, and its application is broadening. In addition to the original work and a recent study highlighted, VSFS has been applied to surfactant-coated particles, biomimetic vesicles, inorganic materials, and emulsion-based colloidal

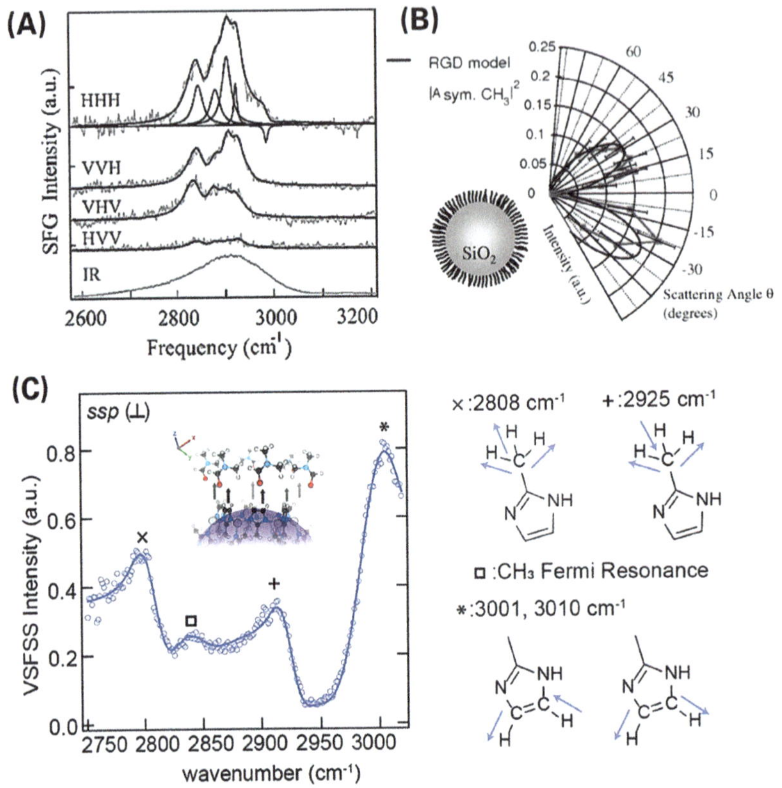

Figure 7.9: (A) VSFS spectra obtained at different polarization conditions. The three-letter codes next to the spectra indicate (H for horizontal, V for vertical) polarizations for SFG, VIS, and IR, respectively. (B) Angular distribution of SFG intensity for the symmetrical CH_3 stretching mode. Reprinted with permission from Roke et al. [158]. (C) VSFS spectrum (SSP polarization) of nanoZIF-8 particles suspended in deuterated-DMF and illustrations of linker vibrational modes corresponding to symbols on spectral peaks. Adapted and reprinted with permission from Mapile et al. [159]. Copyright 2024, American Chemical Society.

systems [159–164]. Liquid-in-liquid emulsions and biological vesicles are not amenable to separation, and it is impracticable to deposit them on a substrate. Therefore, VSFS is a well-suited technique for such soft matters. Generally, adsorption of surfactants enhances the stability of emulsions. Surfactant adsorption has been probed using nonlinear interfacial spectroscopy [134, 165, 166]. The Influence of ions and charged properties of emulsions has also been investigated [118, 126, 167]. Emulsions have gained significant research interest within the interfacial community because this is where hydrophobic oil meets hydrophilic water. Unlike planar oil/water interfaces, they involve curved interfaces. Moreover, questions [168–171] about whether hydroxide versus hydronium ions exhibit greater proclivity for the oil/water interface (both planar and curved) have prompted a vigorous interest. In essence, stabilization mechanisms of emulsions are at the heart of the debate. Unique physical insights into the interfacial structure, charge properties, and stability of colloidal oil/water systems have been gained from VSFS experiments in recent decades [164, 166, 172–180].

As an example, there is a finite volume effect at the nanoscale. VSFS studies [176, 177] of water O/W NEs and W/O (hexadecane) nanodroplets (diam. ~200 nm) involving Span80 as the surfactant revealed that interfacial water molecules have drastically different orientational ordering between the two aqueous interfaces. Based on spectral differences, inside the W/O droplets, water molecules are more strongly hydrogen-bonded compared to interfacial water molecules outside O/W droplets. Water droplets in oil were found to have a much smaller net electrophoretic mobility than oil droplets. Isotope dilution studies also showed that water outside the O/W NE participates significantly in intramolecular coupling, while W/O predominantly exhibits intermolecular coupling. Furthermore, it has been suggested [164] that charge transfer across C–H···O hydrogen bonds stabilizes oil droplets in water. Water at the oil droplet surface revealed a stronger H-bonding network compared with the planar air-water interface.

With the aid of VSFG interfacial spectroscopy, it was found [174] that there is an enhanced bonding between oil and water molecules at the droplet surface relative to that of a planar oil-water interface. This has been shown for low charge nanoemulsions (LCNEs). These bare NEs (size ~300 nm), composed of hexadecane as the oil phase, were prepared in the absence of surfactants. They displayed a low zeta potential (−9.7 ± 5 mV) as a direct result of a stringent multistep cleaning procedure the authors implemented to avoid any potential surface contaminants. These LCNEs were shown to have a measurable free OH vibration, which disappears when surfactants are introduced. The free OH vibration serves as evidence for stronger dispersion interactions between water molecules and the hydrophobic phase of the LCNE. The enhanced bonding is believed to play a role in the droplet stabilization. The picture that emerged is that hexadecane molecules are oriented relatively parallel to the droplet surface in the case of bare NEs. In the presence of sodium dodecyl sulfate (SDS) surfactants, oil molecules reorient to align somewhat vertically along with the surfactant chains via hydrophobic interactions. The interaction of water molecules with the negatively charged surfactant headgroup is more constrained, and the free OH stretch disappears.

In a separate VSFS study [173], it was shown that adsorption of SDS surfactants at the medium-chain triglyceride (MCT) O/W emulsion (diam. >1 μm) interface enhances the interfacial electric field. This manifests in a simultaneous increase in the degree of ordering of the interfacial MCT oil molecules. The evolution of the CH_2/CH_3 peak intensity ratio of MCT at the O/W emulsion interface and the corresponding OH-signal intensities as a function of deuterated-SDS coverage allowed for the determination of the ordering of the MCT molecules. The molecular conformation and magnitude of the interfacial electric field were linked to the emulsion stability. Figure 7.10(A) is a cartoon illustrating the effect of increasing concentration of surfactants on orientational order of MCT molecules, charge of the emulsions, and restructuring of surrounding water molecules. This picture coincides with a decrease in interfacial tension and increased emulsion stability with increasing surfactant concentration. In contrast, for kinetically stable NEs (diam. ~160–260 nm) prepared using SDS, VSFS experiments demonstrated [179] that the SDS alkyl tail is in a disordered state and partially in contact with water. This is in contrast with the traditional view of thermodynamically stable microemulsions. An interfacial structure in which the oil is predominantly oriented parallel with respect to the NE surface has been determined. The alkyl chain of the SDS exhibits few defects to facilitate contact with the oil, whereas the headgroup is surrounded by water molecules. These results suggest that surfactant molecules occupy a large surface area (Figure 7.10(B)).

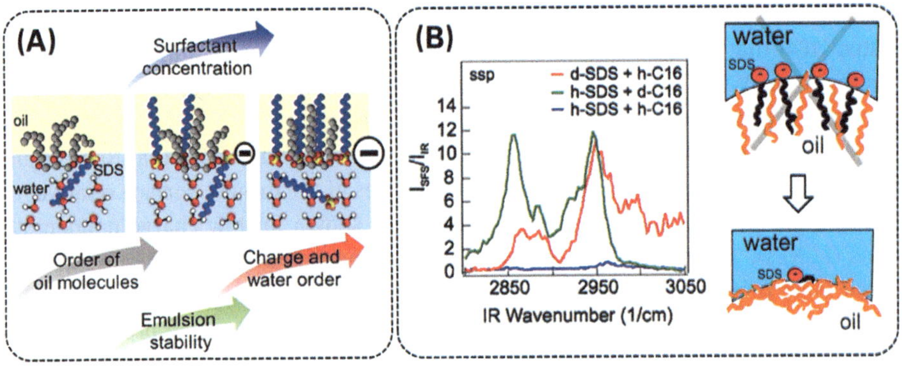

Figure 7.10: Graphical representations of molecular structures at O/W emulsion interfaces as ascertained from VSFS spectral measurements. (A) Enhanced ordering of oil molecules and alignment of water molecules at the O/W emulsion interface as the concentration of anionic surfactant is increased. Reprinted from Hosseinpour et al. [173]. Copyright 2021, The Authors. (B) VSFS spectra showing interference of SDS and hexadecane (C16) molecules. The spectra are collected in the SSP polarization configuration. The green spectrum: droplets prepared with d-C16 oil and h-SDS. The red spectrum: droplets prepared with h-C16 oil and d-SDS. The blue spectrum: droplets prepared with h-C16 oil and h-SDS. The symbols "h" and "d" represent hydrogenated and deuterated, respectively. Based on these data, a cartoon of the correct (in comparison to the traditional view) picture of the SDS and hexadecane at the NE interface is presented. Adapted and reprinted with permission from de Aguiar et al. [179]. Copyright 2011, American Chemical Society.

7.4 Application of $\chi^{(3)}$ Method to Study Colloids

In Section 6.4, we have seen that charged interfaces of colloids affect the nonlinear optical response. The surface charge and the surrounding electric double layer create a noncentrosymmetric environment, which in turn induces a third-order susceptibility, $\chi^{(3)}$, contribution. This response is dependent on electrolyte concentration and the charge density of the colloidal interface. Accordingly, the $\chi^{(3)}$ method has been utilized to decipher acid-base chemistry and the adsorption of charged surfactants at oil/water emulsion interfaces. Figure 7.11(A) shows a titration curve as a function of pH, where the response is the SHS field at a fixed ionic strength. The colloidal system of interest in this study [181] is polystyrene microspheres (diam. ~910 nm) with a carboxylic acid (COOH) surface functional group. As the solution pH increases, deprotonation results in the formation of COO^- ions. Increased charge density results in a greater $\chi^{(3)}$ contribution to the overall SHS signal. Using the Gouy-Chapman model to describe the surface potential and eq. (6.16), the acid-base equilibrium constant (pK_a) of the surface-bound carboxylic group was determined to be 5.4 ± 0.3. Comparing this value to that of similar carboxylic acids dissolved in bulk aqueous solutions, it was revealed that the surface carboxylic groups are more basic. This was attributed to the higher surface free energy of the charged particle-liquid interface.

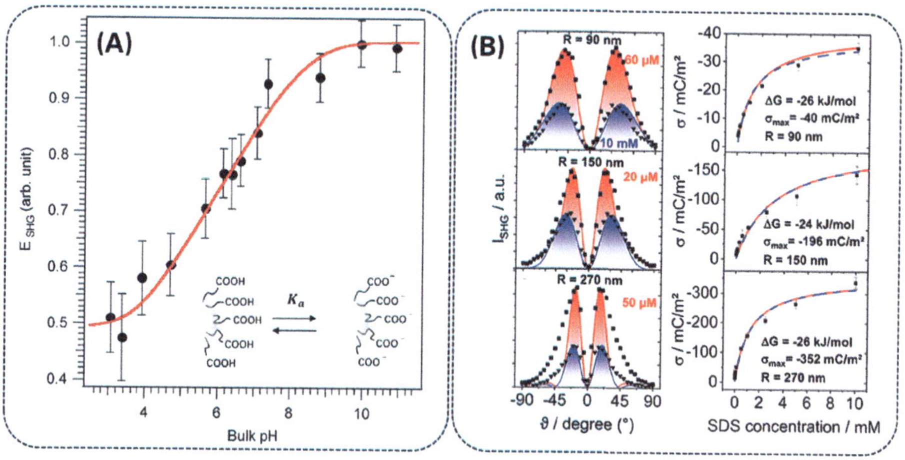

Figure 7.11: (A) Variation of SHS field as a function of solution pH. Adapted and reprinted with permission from Subir et al. [181]. Copyright 2009, American Chemical Society. (B) SHS profiles from hexadecane in water emulsions with mean particle radii of 90, 150, and 270 nm (top to bottom). Square points (red line) and triangles (blue line) represent the lowest and highest SDS concentration, respectively, that was studied for these sizes. Scattering patterns of intermediate concentrations are not shown. Plots of the surface charge density as a function of the SDS concentration for these particles, the Langmuir fit and the modified Langmuir isotherm fit, are shown as red and blue traces, respectively. Adapted and reprinted with permission from Glikman and Braunschweig [182]. Copyright 2021, American Chemical Society.

Recently [182], SHS scattering has been used to study the charge density of NEs as a function of their size. Figure 7.11(B) shows SHS scattering data from SDS surfactant-based NEs of different radii, ranging from 90 to 270 nm. By varying the concentration of SDS and determining an isotherm of charge density as a function of SDS concentration, this study revealed that for sizes >200 nm, the surface charge density is like what can be expected at flat oil/water extended interfaces, while it is considerably reduced when the NE radius is 80 nm. This occurs due to counterion condensation, where the constant charge per particle is maintained by a critical balance between surfactant adsorption and ion condensation, a phenomenon more pronounced at smaller particle sizes. The results highlight that traditional models for flat interfaces are not sufficient to capture these nanoscale effects. Understanding these size-dependent behaviors is crucial for processes involving NEs and their applications in drug delivery, petroleum recovery, and the development of food products and cosmetics.

The samples of SHS, ESFS, and VSFS research work showcased in this chapter highlight the utility of these techniques in elucidating colloidal interfaces. The discussion also shows that extracting information from the SFS light scattered from a colloidal suspension can be convoluted. Correlation between signal intensity and particle number density, interpretation of surface potential, as well as orientational information, are not easily attainable when compared to the use of linear light scattering or planar interfaces. Research is ongoing. Both theoretical and experimental work in this area is pushing the boundaries of understanding colloidal interfaces, and nonlinear light scattering techniques are an integral part of this adventure.

End-of-Chapter Problems

1. Questions related to SFS experimental design and methodology:
   a. How would you design an experiment to confirm that the observed SFS signal from a colloidal solution originates specifically from the isotropic particle surface rather than the surrounding bulk solvent or potential particle aggregation in the bulk? For plasmonic nanoparticles or anisotropic particles, both surface and bulk contributions can generate SFS signals. What experimental approach or analysis method would you use to quantitatively separate these two contributions to accurately study the surface-specific phenomena?
   b. How do we differentiate SHG from two-photon fluorescence?
   c. What control experiments are necessary to ensure that the change in SFS signal over time is due to the intended chemical or structural changes, and not an artifact of the measurement, such as particle settling or laser-induced effects (e.g., heating, photobleaching)?
   d. How could you use polarization-resolved SFS measurements to determine if adsorbed molecules have a preferred orientation (e.g., parallel or perpendicular to the surface normal) and how would you interpret a change in this orientation as a function of surface coverage?
   e. How does the high temporal resolution (using femtosecond pulses) of the SFS method benefit the study of adsorption kinetics compared to other methods?
   f. You are using in situ SHG to monitor the adsorption of a surfactant onto polydisperse polystyrene particles. You observe that the SHG intensity rapidly increases upon the initial addition of the surfactant, and then reaches a plateau.

What specific property of the interface is the SHG signal primarily sensitive to in this scenario (e.g., surface coverage, orientation, charge)?

2. In developing novel organic dyes designed to enhance SHG efficiency, a research team synthesizes three potential candidates (Dye A, Dye B, and Dye C) with varying molecular structures.
   a. Which fundamental molecular property must all three dyes possess to exhibit any nonzero SHG signal?
   b. Dye A is designed with a strong electron-donating group (D) at one end and a strong electron-accepting group (A) at the other end, linked by a conjugated pi system (D-π-A structure). Dye B is a centrosymmetric molecule (it has a center of inversion symmetry). Dye C is like Dye A but has a significantly shorter conjugated bridge. Rank these three dyes (A, B, and C) from highest to lowest expected SHG efficiency and justify your ranking based on the relevant molecular properties.
   c. In a practical application, why might the researchers choose to use Dye A over inorganic SHG crystals (like BBO or KDP) for labeling living cell membranes? Identify one key advantage specific to the use of organic dyes in this context.
   d. Comment on the unique physical and optical properties a dye must possess to be classified as an effective SHG probe.

3. How might a modified Langmuir model be applied to interpret time-dependent SHG scattering intensities to reveal the detailed kinetics and binding constants of molecules adsorbing onto lipid membranes or colloidal surfaces?

4. In an experiment studying silica (SiO_2) colloidal nanoparticles using angle-resolved SHG scattering, researchers find that the SHG signal intensity and polarization change significantly as a function of pH and NaCl concentration. How does changing the pH affect the surface charge of silica particles in water, and consequently, the SHG signal? Simultaneous measurement of the zeta potential shows that it remains relatively constant, even as the SHG signal indicates changes in surface potential. Explain this discrepancy.

5. A researcher is studying the interface of oil droplets in water stabilized by a neutral surfactant, Span80, using VSFS. They observe a strong signal from the C–H stretching region of the surfactant chains at the interface, but no signal from the bulk water (away from the interface). Explain why SFG/SFS is a highly interface-specific technique and why the bulk water does not produce a signal. What two optical processes must a vibrational mode be active in for it to be observable via VSFG? If the researcher uses a different surfactant, SDS, which is an ionic surfac-

tant that creates a charged interface, how might the signal characteristics (e.g., spectral shape, intensity) differ from the neutral Span 80 system?

6. When performing VSFG scattering experiments on colloids dispersed in water, a major challenge is posed by the bulk solvent. What is this challenge, and what experimental strategy is commonly employed to mitigate it, especially when studying O–H stretches?

7. A solvatochromic dye is analyzed in three different solvents, and its maximum absorption wavelengths are measured by UV-Vis spectroscopy:

Solvent A (nonpolar):	**Solvent B (medium polarity)**	**Solvent C (highly polar)**
$\lambda_{max} = 550$ nm	$\lambda_{max} = 515$ nm	$\lambda_{max} = 480$ nm

   a. Is this dye exhibiting positive or negative solvatochromism?
   b. Calculate the transition energies in kcal/mol for each solvent. The formula for converting wavelength to energy is $E = \frac{hc}{\lambda}$.
   c. Explain the underlying physical reason for the observed solvatochromism based on the relative stabilization of the ground and excited states by the solvent polarity.
   d. Assuming this dye adsorbs onto silica particles dispersed in water and yields a λ_{max} of 440 nm. What can be deduced about the interfacial polarity of the colloidal silica/aqueous interface?

8. Progress in science is not without controversy. The topic of interfacial hydroxide ion adsorption [43, 170, 183–188] has been a subject of immense debate for more than a decade. It centers on whether these ions exhibit a specific, counterintuitive affinity for the interface and are responsible for the observed negative zeta potential. There is disagreement between macroscopic experimental results and molecular-level observations, particularly those from nonlinear spectroscopy. Additional contemporary discussions revolve around the detection limits of particles using light-scattering techniques [189–191]. Related articles are cited for further reading.

9. Future of colloidal and interface science is likely to include (a) AI and machine learning for prediction/design, (b) advanced characterization for atomic/molecular-level insights, (c) investigation of complex systems (bio-interfaces, energy materials like MOFs, microplastics), (d) applications in medicine (nanomedicine, gene delivery), smart food and sustainable technology, as well as (e) addressing environmental concerns.
   (A) What innovative in situ characterization tools and methodologies are required to resolve atomic-level transformations of nanomaterials during re-

actions, and how can the data gathered from these advanced techniques be effectively integrated with AI to inform and refine computational models (e.g., DFT and molecular dynamics) that currently make simplifying assumptions about environmental complexity?

(B) What specific interdisciplinary methodologies are needed to combine knowledge of biological systems and engineering principles to create sustainable, bio-inspired materials and designs (e.g., self-cleaning and antifouling surfaces) that effectively address pressing environmental problems?

(C) How can we design smart vesicles (liposomes, emulsions, etc.) for targeted drug and gene delivery that not only exhibit desired release kinetics and cellular absorption pathways but also meet the rigorous regulatory, safety, and manufacturing scalability requirements for widespread clinical application?

(D) Microplastics are an emerging concern. How can colloidal science research move beyond detection and characterization to develop effective, large-scale remediation strategies or design new, truly sustainable materials that do not pose similar long-term environmental risks?

Epilogue

Colloidal science and interfacial chemistry involve the study of systems where matter is finely dispersed and interacts across boundaries. Throughout history, people have intuitively manipulated matter at the microscopic scale. They produced vibrant paints, durable ceramics, and early forms of medicine without knowing the underlying principles of colloids. In fact, it has been an arduous path to recognize colloids as a unique class of matter and establish their study as a true interdisciplinary science. In its nascent years, colloidal investigations relied on observation and classification, employing simple optical methods such as microscopes to visualize these particles. Within half a century, the advent of lasers and robust foundations of light scattering theories have revolutionized the study of colloids. Characterization of size and charge properties of colloids based on light scattering methods is nowadays routine. Advanced laser techniques to track, design, and manipulate colloids are at the forefront of engineering at the microscale.

These days, the fields of nonlinear spectroscopy and colloids have converged in a powerful synergy. The utility of sum frequency scattering to extract information from colloidal interfaces has blossomed. The future of colloidal science is set to be transformative. The research highlighted throughout Chapter 7 demonstrates the power of SFS for studying colloidal interfaces. Pioneering fundamental research discussed makes these in situ and time-resolved characterizations possible. The nonlinear light scattering technique has a lot of advantages. Its inherent interfacial specificity, spectroscopic nature that provides molecular-level information (order, orientation, and even chirality), and sensitivity to charged interfaces are some of the major attributes that make SFS an excellent laser-based technique to study chemical interfaces of colloids. However, as with any analytical tool, there are limitations and challenges. Thus, the future NLS work in this area will not only involve applying this technique to a wide range of colloids but also improving its analytical and data interpretation capabilities so that it can be applied to more complicated colloidal systems.

Challenges are generally associated with particle properties, signal and sensitivity issues, and data interpretation complexity. For example, as of now, the SFS technique is limited to spherical particles. Mie scattering and similar theories are not always applicable in analyzing non-spherical particles or particles with heterogeneous size distributions. The scattering patterns measured using AR-SFS are likely to be different for colloids of different shapes and dispersity. Thus, future research is likely to involve extending light scattering theories [192] involving nonspherical particles and heterogeneous colloids to SFS. Furthermore, unless resonantly enhanced, the SHS efficiency is much lower compared to fluorescence and other linear scattering responses. Given that lasers are often limited to a narrow wavelength range, interfacial adsorption studies have predominantly been conducted using SHG dye molecules that are resonant with these specific lasers. ESFG, which involves light sources with broad

 | https://doi.org/10.1515/9783111018386-008

wavelength and greater tunability range, has been applied to planar interfaces. Its application to colloidal systems will make it possible to probe electronic transitions of a diverse group of organic compounds. Advancing existing SFS methodology to probe heterogeneous colloidal suspensions, as well as a broader group of organic molecules, will have an immense impact on understanding molecular interactions of colloids in the realm of biological, medical, and environmental sciences.

Another challenge associated with SFS is related to the interference effects. As noted in Chapter 6, signals originating from the surrounding medium (e.g., $\chi^{(3)}$ contribution), as well as the bulk of the particle, can interfere with those stemming from the interface. This can convolute physical insights into interfacial processes and structural properties. Incorporation of phase-sensitive (PS) heterodyne detection [193, 194] in SFG has been a significant advancement in this front. Phase-resolved measurements provide improved analysis of the interfacial structure. It is anticipated that PS-SFS measurements will become prominent in the near future, which will facilitate a better understanding of charged-colloidal interfaces. Implementation of this technique to study depth-dependent interfacial structural anisotropy is emerging [195]. Investigations into polyelectrolytes and other soft colloids, which usually have a thick interfacial layer, will thus be the next frontier to conquer using this advanced methodology. Another area of focus will be interfacial ultrafast dynamics at colloidal interfaces. Pump-probe SFG has been applied to planar interfaces, and its extension to colloids is only natural [196–199]. Furthermore, the successful implementation of two-dimensional electronic-vibrational sum frequency spectroscopy (2D-EVSFG) to study planar interfaces has been demonstrated [200, 201]. Deciphering electronic-vibrational couplings is essential in the elucidation of interfacial processes such as charge, energy, and proton transfer. Plausible application of 2D-EVSFG to colloidal interfaces can have promising implications in the realm of nanoscience.

The exemplary research showcased demonstrates the versatility of SFS. Yet, there are a multitude of different types of colloids and pertinent interfacial properties that require further exploration. Some examples have been noted, but the scope is wide open. In the areas of biology and medical science, the study of protein adsorption and conformation at interfaces is vital for understanding phenomena like bio-adhesion, biocompatibility of medical implants, and marine biofouling. Investigation of molecular adsorption from the gas phase to aerosol particles is critical in atmospheric science. VSFS can provide a wealth of information on water structure and ion behavior at the surface of colloidal primary and secondary aerosols. Within the aquatic environment, there are various colloidal substances, including mineral oxides, particulate natural organic matter, and contaminants such as nanomaterials and colloidal microplastics. Understanding chemical processes at these colloidal interfaces is also at the forefront of SFS studies. We have also seen the application of SFS to the study of liquid-in-liquid colloids. Further insights into emulsion stability, drug-delivery mechanisms using biomimetic vesicles, and long-range stability of soft colloids can be obtained with the aid of SFS and complementary light scattering techniques. Lastly, while the focus has been on the application of SFS to study colloidal interfaces, understanding the nonlinear optical properties of nanocolloids is also in demand. Engineering nanomaterials in a colloidal state for efficient NLO response is an active area of research.

The ability to characterize colloidal interfaces in situ and nondestructively, without labels, provides a unique opportunity to understand interfacial properties of colloids using SFS. Colloidal and interfacial science is truly interdisciplinary. It is an area ripe for further exploration based on nonlinear light scattering methods elaborated herein and awaits new discoveries.

Glossary

Adhesion – The process of attachment of interfaces between phases or components that is maintained by intermolecular forces, chain entanglements, or both.

Adsorbate – A molecular species of gas, dissolved substance, or liquid that adheres to or is adsorbed in an extremely thin surface layer of a solid substance.

Adsorbent – A solid surface or the surface of a condensed phase where adsorption occurs.

Adsorption – An increase in the concentration of a dissolved substance at the interface of a condensed and a liquid phase due to the operation of surface forces. Adsorption can also occur at the interface of a condensed and a gaseous phase.

Adsorption isotherm – A plot generated in a typical adsorption experiment that describes the relationship between the amount adsorbed and the concentration of the adsorptive at an equilibrium under constant temperature.

Adsorptive – The material that is present in one or other (or both) of the bulk phases and capable of being adsorbed.

Anisotropic – Having properties that depend on direction.

Amphiphilic – A term used to describe a compound containing a large organic cation or anion that possesses a long unbranched hydrocarbon.

Aerosol – A sol in which the dispersed phase is a solid, a liquid, or a mixture of both and the continuous phase is a gas (usually air).

Agglomeration – A process in which dispersed molecules or particles assemble rather than remain as isolated single molecules or particles. It is a reversible process.

Aggregation – A process whereby dispersed molecules or particles form aggregates. It is an irreversible process.

Bathochromic shift – Shift of a spectral band to lower frequencies (longer wavelengths) owing to the influence of substitution or a change in environment. It is informally referred to as a red shift.

Brownian motion – The movement of particles in a colloidal system caused by collision with the molecules in the fluid in which the particles are imbedded.

Capillary electrophoresis – A separation method based solely on the differences in the electrophoretic mobilities of charged species in capillaries.

Centrosymmetric – An object that is symmetric with respect to a central point. In crystallography, it refers to a material assigned to a point group that includes an inversion symmetry.

Chemisorption – Adsorption which results from chemical bond formation (strong interaction) between the adsorbent and the adsorbate in a monolayer on the surface.

Coagulation – When a sol is colloidally unstable (i.e., the rate of aggregation is not negligible), the formation of aggregates is called coagulation or flocculation.

Coalescence – The disappearance of the boundary between two particles (usually droplets or bubbles) in contact, or between one of these and a bulk phase, followed by changes of shape leading to a reduction of the total surface area. The flocculation of an emulsion, viz., the formation of aggregates, may be followed by coalescence.

 | https://doi.org/10.1515/9783111018386-009

Coherence length – The distance over which order is maintained. The propagation distance over which an electromagnetic wave can travel without losing a specified degree of coherence, that is, propagate without exhibiting a phase discontinuity. In the context of SHG, it is the maximum length within which frequency doubling is efficient.

Coherent process – Pertains to a fixed phase relationship between electric fields at different locations or times. Light arising from different scattering centers is said to be coherent when the phase difference between their waves is constant.

Colloids (or colloidal system) – The term refers to a state of subdivision, implying that the molecules or polymolecular particles dispersed in a medium have at least in one direction a dimension roughly between 1 nm and 1 µm, or that in the system discontinuities are found at distances of that order.

Contact angle – When a liquid does not spread on a solid substrate, a contact angle (θ) is formed between two of the interfaces at the three-phase line of contact. It must always be stated which interfaces are used to define θ. It is often necessary to distinguish between the advancing contact angle, θ_a, the receding contact angle, θ_r, and the equilibrium contact angle, θ_e. When $\theta_a \neq \theta_r$, the system is said to exhibit contact angle hysteresis.

Critical micelle concentration (CMC) – There is a relatively small range of concentrations separating the limit below which virtually no micelles are detected and the limit above which virtually all additional surfactant molecules form micelles.

Debye length – Within the context of charged surfaces in an electrolyte solution, it is a measure of the distance over which the electrostatic field is reduced to $\frac{1}{e}$. It represents the thickness of the electric double layer.

Desorption – The converse of adsorption, i.e., the decrease in the amount of adsorbed substance.

Diffraction – A modification which light (or electron, neutron beams, etc.) undergoes in passing by the edges of opaque bodies or through narrow slits or in being reflected from ruled surfaces (or crystalline materials). The light waves, owing to their wave-like nature, appear to be deflected and produce fringes of parallel light and dark bands corresponding to regions of constructive reinforcement or destructive interference, respectively, of the waves.

Diffusion – The spreading or scattering of a gaseous or liquid material. Eddy diffusion in the atmosphere is the process of transport of gases due to turbulent mixing in the presence of a composition gradient. Molecular diffusion is the net transport of molecules, which results from their molecular motions alone in the absence of turbulent mixing; it occurs when the concentration gradient of a particular gas in a mixture differs from its equilibrium value. Eddy diffusion is the most important mixing process in the lower atmosphere, while molecular diffusion becomes significant at the lower pressures of the upper atmosphere.

Dissolved organic carbon – Concentration of carbon found dissolved in water samples from aquatic systems, measured as total elemental carbon.

Dissolved organic matter – Analogous to dissolved organic carbon, but refers to the entire organic pool dissolved in water. Operationally defined as the "dissolved" portion of organic carbon in a liquid sample that passes through a filter with a pore size of 0.7 μm and 0.22 μm. This operational definition stems from the practical need to separate dissolved from particulate components and is crucial for studying the complex, heterogeneous mixture of organic compounds that includes carbohydrates, proteins, amino acids, and humic substances.

DLVO theory – Named after Derjaguin, Landau, Verwey, and Overbeek, it is a theoretical framework used to explain the stability of lyophobic colloids.

Electric double layer – The Coulombic interaction of interfacial charges (e.g., ions) and the magnetic or electrostatic interaction of interfacial molecules lead to particularly complex interfacial structures. Complex interfacial profiles that can be approximated by two distinct sub-layers with different physical properties (e.g., structure and/or nature and/or composition) are referred to as interfacial double-layers. Examples of such approximated complex profiles are: the electrical double-layer consisting of a surface charge layer (i.e., a two-dimensional distribution of one type of ions) and a diffuse charge layer (counterions distributed over the space region next to the surface); the approximated profile of the orientation angle of anisotropic liquid molecules within a "double-layer" consisting of a distribution of so-called anchored molecules which are perturbed (strongly bound and orientated) by the surface, and the adjacent, so-called, transition layer, i.e., the region where the surface perturbation is damped.

Electroosmosis – Motion of a liquid through a capillary (or membrane or plug) because of the application of an electric field across the capillary (or membrane or plug).

Electrophoresis – The motion of colloidal particles in an electric field.

Emulsions – A fluid colloidal system in which liquid droplets and/or liquid crystals are dispersed in a liquid. The droplets often exceed the usual limits for colloids in size. An emulsion is denoted by the symbol O/W if the continuous phase is an aqueous solution and by W/O if the continuous phase is an organic liquid (an "oil"). More complicated emulsions, such as O/W/O (i.e., oil droplets contained within aqueous droplets dispersed in a continuous oil phase), are also possible.

Emerging contaminants – Chemical substances that show potential significant risks to human health and the environment. Contaminants of emerging concern can include nanoparticles, pharmaceuticals, personal care products, estrogen-like compounds, flame retardants, detergents, and some industrial chemicals.

Fate and transport – The behavior, movement, and eventual disposition of chemicals within the environment. It involves both physical and chemical processes by which a contaminant enters the environment, is transformed, and is relocated.

Flocculation – See coagulation.

Foam – A dispersion in which a large proportion of gas by volume in the form of gas bubbles is dispersed in a liquid, solid, or gel. The diameter of the bubbles is usually larger than 1 μm, but the thickness of the lamellae between the bubbles is often in the usual colloidal size range. The term froth has been used interchangeably with foam. Froth may be distinguished from foam by the fact that the former is stabilized by solid particles (as in froth flotation) and the latter by soluble substances.

Gel – Nonfluid colloidal network or polymer network that is expanded throughout its whole volume by a fluid.

Gibbs' monolayer – A soluble monolayer formed at an interface by a compound that exhibits partitioning into at least one of the phases separated by the interface.

Heterogeneous – A mixture that is nonuniform in nature. A chemical system that is composed of different substances or two or more phases of the same substance.

Homogeneous – A substance that is uniform in nature. A chemical system is homogeneous if each intensive macroscopic property is constant throughout the system. A particle that is spatially uniform with respect to the chemical composition of the constituent polymer(s).

Humic substance – Nonvolatile organic anionic polyelectrolytes, of biological origin, with complex structures and variable compositions. They occur naturally as deposits on sediment and soil particles and constitute the dissolved organic carbon in natural waters. Humic substances are classified according to solubility and may contain humic acid, fulvic acid, and humin.

Hypsochromic shift – Shift of a spectral band to a higher frequency or shorter wavelength upon substitution or change in medium. It is informally referred to as a blue shift.

Inhomogeneous – Something that is not homogeneous.

Interface – The plane ideally marking the boundary between two phases. If one of the two phases is gaseous, it is referred to as a surface.

Interfacial excess – In relation to the concept of surface excess, this quantity describes the surface concentration of a component. It is usually represented as molecules per unit area.

Interfacial layer – The inhomogeneous space region intermediate between two bulk phases in contact, and where properties are significantly different from, but related to, the properties of the bulk phases. Examples of such properties are: compositions, molecular density, orientation or conformation, charge density, pressure tensor, electron density, etc. The interfacial properties vary in the direction normal to the surface. Complex profiles of interfacial properties occur in the case of multicomponent systems with coexisting bulk phases, where attractive/repulsive molecular interactions involve adsorption or depletion of one or several components. This interfacial region may also be regarded as a distinct, though not autonomous, phase and be called the interphase.

Interfacial tension – See surface tension.

Interphase – See interfacial layer

Isotropic – A physical property (e.g., optical, magnetic, mechanical) of a material that is independent of direction. Isotropic materials exhibit uniformity in all orientations.

Langmuir monolayer – An insoluble monolayer formed by a layer of amphiphilic molecules at the interface of air and water. See "Monolayer."

LASER – A source of ultraviolet, visible, or infrared radiation which produces light amplification by stimulated emission of radiation from which the acronym is derived. The light emitted is coherent except for super-radiance emission. The essential elements of a laser are: 1, an active medium; 2, a pumping process to make a population inversion; and 3, a suitable geometry of optical feedback elements. The active medium consists of a host material (gas, liquid, or solid) containing an active species.

Liposome – Originally a lipid droplet in the endoplasmic reticulum of a fatty liver. Now applied to an artificially formed lipid droplet, small enough to form a relatively stable suspension in aqueous media, and with potential use in drug delivery. In general, liposomes are microscopic in size and contain a lipid bilayer mimicking a natural biological cell membrane.

Lyophilic – A substance that has a strong affinity or preference for a specific solvent medium, meaning "solvent-loving," such as water-loving (hydrophilic) or fat-loving (lipophilic). The substances form stable colloids easily by simple mixing due to strong interactions.

Lyophobic – Molecular entities (or parts of them) that are "solvent-rejecting" or have little to no affinity for the dispersion medium. They do not naturally mix or attract, leading to unstable colloids requiring special preparation and stabilization, such as with electric charges, to prevent coagulation.

MASER – Microwave Amplification by Stimulated Emission of Radiation.

Macromolecule – A molecule of high relative molecular mass, the structure of which essentially comprises the multiple repetition of units derived, actually or conceptually, from molecules of low relative molecular mass.

Micelle – Surfactants in solution are often association colloids, that is, they tend to form aggregates of colloidal dimensions, which exist in equilibrium with the molecules or ions from which they are formed. Such aggregates are termed micelles.

Microgel – A particle of gel of any shape with an equivalent diameter of approximately 0.1–100 μm.

Mie scattering – The scattering of electromagnetic radiation by spherical particles of any size *r*, relative to the wavelength, λ. Since the cases $r \ll \lambda$ and $r \gg \lambda$ are covered by Rayleigh (dipole) scattering and geometric scattering theories, respectively, Mie scattering often refers to the case of $r \approx \lambda$.

Monolayer – A single, closely packed layer of atoms or molecules. The term "floating monolayer" is used for certain spread monolayers or films which are in the condensed – often solid – state. The term "Langmuir monolayer" has recently been coined for spread monolayers. This usage is not recommended.

Natural organic matter – Organic material in natural environments, like surface and ground water, derived from plants and animals. It is a complex mixture of humic and nonhumic substances.

Noncentrosymmetric – A substance that does not have points of inversion symmetry throughout its volume.

Nonlinear optics – A branch of optics that studies phenomena where a material's response to an applied electromagnetic field is not linearly proportional to the intensity of the applied field.

Parametric process – A light-matter interaction process in which there is no net transfer of energy or momentum from the electromagnetic field to the physical system. That is, upon interaction, the quantum state of the matter remains unchanged.

Particulate organic matter – Operationally defined as the fraction of total organic matter that does not pass through a filter. Generally, this represents material that is collected on the filter and can range from 0.053 mm to 2 mm in soil, or greater than 0.2–0.7 μm in water.

Phase matching – The condition that must be met for efficient generation of most nonlinear optical responses. In considering the interaction of two electromagnetic waves of different frequencies, as in the case of SHG, the process is efficient if the phase velocities of the waves are equal. In other words, phase matching is a consequence of conservation of momentum.

Physisorption – A process in which the forces involved are intermolecular forces (van der Waals forces) of the same kind as those responsible for the imperfection of real gases and the condensation of vapors, and which do not involve a significant change in the electronic orbital patterns of the species involved. The term "van der Waals adsorption" is synonymous with physical adsorption, but its use is not recommended.

Polymer – A substance composed of macromolecules.

Porosity – A concept related to texture, referring to the pore space in a material.

Primary aerosol – Aerosol generated by a nebulizer. Usually, primary aerosols contain coarse droplets, they are polydisperse in terms of drop diameters, and they are turbulent.

Rayleigh scattering – The scattering of light by particles and molecules that are much smaller than the wavelength of the light. In the ideal case, the process is one of pure dipole and induced dipole interactions with the electric field of the light wave. The scattering cross section for light of wavelength λ is proportional to λ^{-4}. The phenomenon is named in honor of Lord Rayleigh.

Rayleigh ratio – The quantity used to characterize the scattered intensity at the scattering angle, θ.

Raman scattering – Discovered by Sir C.V. Raman in 1928, it describes the inelastic scattering of light by matter. In vibrational Raman spectroscopy, the energy of the scattered photon is altered due to the vibrational modes of the interacting molecule. Compared to the energy of the incident photon, the scattered photon can either be lower (Stokes Raman) or higher (anti-Stokes Raman).

Resonance enhancement – A boost in optical response (e.g., increased intensity) that occurs when the frequency of the electronic transition matches that of the incident laser field. In second-order processes, resonance enhancement also occurs if the frequency of the SHG or SFG light field overlaps with that of the electronic transition.

Second harmonic generation – Conversion of the input monochromatic radiation beam into a double frequency (half wavelength) beam.

Secondary organic aerosol – Airborne particles comprised of organic compounds that have undergone atmospheric chemical transformation. They constitute a major component of tropospheric and consist of a mixture of compounds, including VOCs, nitrates, and sulfates.

Sedimentation – Separation of a dispersed system under the action of a gravitational or centrifugal field according to the different densities of the components.

Sedimentation coefficient – Velocity of sedimentation divided by the acceleration of the force field (gravitation or centrifugation).

Sedimentation velocity – Velocity of a component in a fluid relative to the fluid in the direction of gravitational or centrifugal acceleration. Also called the sedimentation rate.

Sol – A fluid colloidal system of two or more components, e.g., a protein sol, a gold sol, an emulsion, a surfactant solution above the critical micelle concentration.

Sorption – The process by which a substance (sorbate) is sorbed (adsorbed or absorbed) on or in another substance (sorbent). Adsorption is a surface phenomenon, whereas absorption refers to the uptake of a substance into the bulk volume of another substance.

Spectrometry – The measurement of radiation upon interaction of an electromagnetic wave with matter as a means of obtaining information about the systems and their components.

Spectroscopy – The study of physical systems by the electromagnetic radiation with which they interact or that they produce.

Substrate – A chemical species, the reaction of which with some other chemical reagent is under observation. In surface chemistry, the term refers to the surface (usually a solid surface) on which the adsorption takes place. The term should be used with care, as it has other meanings in catalysis and in film science and technology. Either the context or a specific statement should always make it clear which chemical species in a reaction is regarded as the substrate.

Sum frequency generation – A nonlinear optical phenomenon where two input laser beams of different frequencies interact within a non-centrosymmetric medium to generate an output beam at a new frequency that is the sum of the two input frequencies.

Surface coverage (θ) – the number of molecules on a surface divided by the maximum number of molecules that form a saturated or complete monolayer on that surface. A complete monolayer is a closely packed single layer of the adsorbate.

Surface excess – According to IUPAC, it is the difference between the total moles of a substance in each system and the amount that would be present in the bulk adjoining phases up to a hypothetical geometrical plane (e.g., such as the Gibbs dividing surface) separating the phases.

Surface plasmon – A coherent, collective electron oscillation that propagates along the interface between a metal and a dielectric medium.

Surface tension – Work required to increase the surface area divided by that area. When two phases are non-gaseous, it is often called interfacial tension.

Surfactant – A substance that lowers the surface tension of the medium in which it is dissolved, and/or the interfacial tension with other phases, and, accordingly, is positively adsorbed at the liquid/vapor and/or at other interfaces.

Tensor – A mathematical structure (such as an array of numbers or functions) that can interact with and transform other mathematical entities in accordance with a set of transformation rules. The order (or rank) of a tensor is the number of dimensions needed to describe this algebraic entity.

Tyndall effect – The scattering of a beam of light by particles in a colloidal suspension, making the path of the light visible.

Ultrafast dynamics – Molecular events or processes that occur at very short timescales (approximately in the range of pico- to femtoseconds or shorter).

Virtual state – A short-lived, unobservable quantum state that is not an eigenstate of the Hamiltonian. It does not correspond to a well-defined value of energy. It is produced when the electron cloud interacts with an electromagnetic field. The resulting field-induced polarization and distortion of electrons is described as an intermediate superposition or mixing of states, i.e., a virtual state.

Zeta potential – Potential at the plane where slip with respect to the bulk solution is postulated to occur. It is the electrokinetic potential at the "slipping plane," which is the boundary separating the mobile fluid layer surrounding a particle from the bulk fluid.

Zimm plot – A diagrammatic representation of data on scattering from large particles, used for the simultaneous evaluation of the mass average molar mass, the second virial coefficient, and the *z*-average radius of gyration.

Source: IUPAC Compendium of Chemical Terminology, 3rd ed. International Union of Pure and Applied Chemistry; 2006. Online version 3.0.1, 2019.

Fundamental Constants

Avogadro's number, $N_A = 6.022 \times 10^{23}\ \mathrm{mol}^{-1}$
Boltzmann Constant: $k_B = 1.38065 \times 10^{-23}\ \frac{\mathrm{J}}{\mathrm{K}}$
Electron rest mass: $m_e = 9.109 \times 10^{-31}\ \mathrm{kg}$
Elementary charge: $e = 1.602 \times 10^{-19}\ \mathrm{C}$
Faraday constant: $F = 9.648 \times 10^{4}\ \frac{\mathrm{C}}{\mathrm{mol}}$
Gas Constant: $R = 0.0821\ \mathrm{L \cdot atm/mol \cdot K}$; $R = 8.3145\ \frac{\mathrm{J}}{\mathrm{mol\ K}} = 8.3145\ \frac{\mathrm{m^3\ Pa}}{\mathrm{mol\ K}}$
Gravitational constant: $G = 6.673 \times 10^{-11}\ \mathrm{m^3/(kg \cdot s^2)}$
Magnetic permeability of free space: $\mu_0 = 4\pi \times 10^{-7}\ \mathrm{H/m}$
Neutron rest mass: $m_n = 1.6749 \times 10^{-27}\ \mathrm{kg}$
Permittivity of vacuum: $\varepsilon_0 = 8.854 \times 10^{-12}\ \frac{\mathrm{C^2}}{\mathrm{J \cdot m}}$
Proton rest mass: $m_p = 1.6726 \times 10^{-27}\ \mathrm{kg}$
Planck's constant: $h = 6.626 \times 10^{-34}\ \mathrm{J \cdot s}$; reduced Planck's constant $\hbar = 1.055 \times 10^{-34}\ \mathrm{J \cdot s}$
Speed of light: $c = 2.998 \times 10^{8}\ \mathrm{m/s}$
Standard acceleration of gravity: $g = 9.807\ \mathrm{m/s^2}$

 | https://doi.org/10.1515/9783111018386-010

Common Conversions

$1\ \text{amu} = 1.661 \times 10^{-27}\text{kg}$

$1\ \text{cal} = 4.184\ \text{J} = 4.184 \times 10^{-7}\ \text{erg}$

$1\ \text{dyne} = 10^{-5}\ \text{N}$

$1\ \text{atm} = 760\ \text{Torr} = 760\ \text{mmHg} = 101,325\ \text{Pa}$

$1\ \text{eV} = 1.602 \times 10^{-19}\ \text{J} = 8065.73\ \text{cm}^{-1} = 2.418 \times 10^{14}\ \text{Hz} = 23.061 \frac{\text{kcal}}{\text{mol}} = 96.487 \frac{\text{kJ}}{\text{mol}}$

$1\ \text{D} = 3.336 \times 10^{-30}\ \text{C} \cdot \text{m}$

$1\ \text{esu} = 3.336 \times 10^{-10}\ \text{C}$

$1\ \text{L} = 10^{3}\ \text{mL} = 10^{3}\ \text{cm}^{3} = 10^{-3}\ \text{m}^{3} = 1\ \text{dm}^{3}$

$1\text{Å} = 10^{-10}\ \text{m} = 10^{-1}\ \text{nm}$

 | https://doi.org/10.1515/9783111018386-011

References

[1] Docter D, Westmeier D, Markiewicz M, Stolte S, Knauer SK, Stauber RH. The nanoparticle biomolecule corona: Lessons learned – Challenge accepted?. Chemical Society Reviews. 2015;44 (17):6094–121.

[2] Patel N, Guillemette R, Lal R, Azam F. Bacterial surface interactions with organic colloidal particles: Nanoscale hotspots of organic matter in the ocean. Plos One. 2022;17(8):e0272329.

[3] Yan J, Manelski R, Vasilas B, Jin Y. Mobile colloidal organic carbon: An underestimated carbon pool in global carbon cycles?. Frontiers in Environmental Science. 2018;6. 1–12.

[4] Tiller CL, O'Melia CR. Natural Organic Matter and Colloidal Stability: Models and Measurements. In: Tadros TF, Gregory J, editors. Colloids in the Aquatic Environment. Oxford: Elsevier; 1993. pp. 89–102.

[5] Philippe A, Schaumann GE. Interactions of dissolved organic matter with natural and engineered inorganic colloids: A review. Environmental Science & Technology. 2014;48(16):8946–62.

[6] Angelico R, Colombo C, Di Iorio E, Brtnický M, Fojt J, Conte P. Humic substances: From supramolecular aggregation to fractal conformation – Is there time for a new paradigm?. Applied Sciences. 2023;13(4):2236.

[7] Allen S, Allen D, Karbalaei S, Maselli V, Walker TR. Micro(nano)plastics sources, fate, and effects: What we know after ten years of research. Journal of Hazardous Materials Advances. 2022;6:100057.

[8] Peller JR, Nelson CR, Babu BG, Iceman C, Kostelnik E. A review of microplastics in freshwater environments: Locations, methods, and pollution loads. Contaminants in Our Water: Identification and Remediation Methods. ACS Symposium Series. 1352: American Chemical Society. 2020;65–90.

[9] Iroegbu AOC, Ray SS, Mbarane V, Bordado JC, Sardinha JP. Plastic pollution: A perspective on matters arising: Challenges and opportunities. ACS Omega. 2021;6(30):19343–55.

[10] Al Harraq A, Bharti B. Microplastics through the lens of colloid science. ACS Environmental Au. 2022;2(1):3–10.

[11] Tadros TF. Colloids in paints. Colloids in Paints. 2010;1–9.

[12] Mahmood Q, Khan AF, Khan A. Chapter 25 – Colloids in the Environmental Protection – Current and Future Trends. In: Fanun M, editor. The Role of Colloidal Systems in Environmental Protection. Amsterdam: Elsevier; 2014. pp. 635–77.

[13] Harman CLG, Patel MA, Guldin S, Davies G-L. Recent developments in Pickering emulsions for biomedical applications. Current Opinion in Colloid & Interface Science. 2019;39:173–89.

[14] Zhou Y, Yin D, Chen W, Liu B, Zhang X. A comprehensive review of emulsion and its field application for enhanced oil recovery. Energy Science & Engineering. 2019;7(4):1046–58.

[15] Gbadamosi AO, Junin R, Manan MA, Agi A, Yusuff AS. An overview of chemical enhanced oil recovery: Recent advances and prospects. International Nano Letters. 2019;9(3):171–202.

[16] Brown P, Butts CP, Cheng J, Eastoe J, Russell CA, Smith GN. Magnetic emulsions with responsive surfactants. Soft Matter. 2012;8(29):7545–46.

[17] Qiao K, Tian W, Bai J, Wang L, Zhao J, Du Z, et al. Application of magnetic adsorbents based on iron oxide nanoparticles for oil spill remediation: A review. Journal of the Taiwan Institute of Chemical Engineers. 2019;97:227–36.

[18] Ridi F, Fratini E, Baglioni P. Cement: A two thousand year old nano-colloid. Journal of Colloid and Interface Science. 2011;357(2):255–64.

[19] Ottewill RH, Sing KSW, Bensted J. Chemistry of colloidal silicates and cements [and Discussion]. Philosophical Transactions of the Royal Society of London Series A, Mathematical and Physical Sciences. 1983;310(1511):67–78.

[20] Turkevich J. Colloidal gold. Part I. Gold Bulletin. 1985;18(3):86–91.

[21] Freestone I, Meeks N, Sax M, Higgitt C. The lycurgus cup – A roman nanotechnology. Gold Bulletin. 2007;40(4):270–77.

 | https://doi.org/10.1515/9783111018386-012

[22] Hauser EA. The history of colloid science: In memory of Wolfgang Ostwald. Journal of Chemical Education. 1955;32(1):2.

[23] Lyklema J. Colloidal models. A bit of history. Journal of Colloid and Interface Science. 2015;446:308–16.

[24] Brown R. A brief account of microscopical observations on the particles contained in the pollen of plants and the general existence of active molecules in organic and inorganic bodies. Edinburgh New Philosophical Journal. 1828(4);358–71.

[25] Brown R. The Miscellaneous Botanical Works of Robert Brown, Vol. 1. London: R Hardwicke; 1866.

[26] Clayton W. The foundations of colloid chemistry: A selection of early papers bearing on the subject. Nature. 1925;116(2918):494–95.

[27] Hatschek E, editor. The Foundations of Colloid Chemistry. A Selection of Early Papers on the Subject. New York: The Macmillan Company; 1925.

[28] Wisniak J. Thomas Graham. I. Contributions to thermodynamics, chemistry, and the occlusion of gases. Educación Química. 2013;24(3):316–25.

[29] Mokrushin SG. Thomas Graham and the definition of colloids. Nature. 1962;195(4844):861–.

[30] Graham TX. Liquid diffusion applied to analysis. Philosophical Transactions of the Royal Society of London. 1861;151:183–224.

[31] Faraday MX. The Bakerian Lecture. – Experimental relations of gold (and other metals) to light. Philosophical Transactions of the Royal Society of London. 1857;147:145–81.

[32] Ostwald CWW. An Introduction to Theoretical and Applied Colloid Chemistry, “The World of Neglected Dimensions”. New York: John Wiley & Sons, Inc.; 1922.

[33] Broadbent TA, Einstein AB, Hertz H, Dryden HL, Murnaghan FP, Bateman H. Investigations on the theory of the Brownian movement. The Mathematical Gazette. 1928;41:231–.

[34] Bigg C. Evident atoms: Visuality in Jean Perrin’s Brownian motion research. Studies in History and Philosophy of Science Part A. 2008;39(3):312–22.

[35] Royall CP, Poon WCK, Weeks ER. In search of colloidal hard spheres. Soft Matter. 2013;9(1):17–27.

[36] Poon WCK, Weeks ER, Royall CP. On measuring colloidal volume fractions. Soft Matter. 2012;8 (1):21–30.

[37] Vlassopoulos D, Cloitre M. Tunable rheology of dense soft deformable colloids. Current Opinion in Colloid & Interface Science. 2014;19(6):561–74.

[38] Cosgrove T, editor. Colloid Science Principles, Methods and Applications, 2nd ed. Bristol, UK: John Wiley and Sons, Ltd.; 2010.

[39] Hao T, Riman RE. Calculation of interparticle spacing in colloidal systems. Journal of Colloid and Interface Science. 2006;297(1):374–77.

[40] Jasper JJ. The surface tension of pure liquid compounds. Journal of Physical and Chemical Reference Data. 1972;1(4):841–1010.

[41] Adamson AW. Physical Chemistry of Surfaces. New York: John Wiley & Sons, Inc.; 1990.

[42] Barnes G, Gentle I. Interfacial Science: An Introduction, 2nd ed. Oxford: Oxford University Press; 2011.

[43] Jungwirth P, Winter B. Ions at aqueous interfaces: From water surface to hydrated proteins. Annual Review of Physical Chemistry. 2008;59(1):343–66.

[44] Tian J, He Y, Li F, Peng W, He Y. A comprehensive review on the mechanism of contact electrification. Journal of Materials Chemistry A. 2025;13(4):2505–36.

[45] McCarty LS, Winkleman A, Whitesides GM. Ionic electrets: Electrostatic charging of surfaces by transferring mobile ions upon contact. Journal of the American Chemical Society. 2007;129 (13):4075–88.

[46] Sotthewes K, Gardeniers HJGE, Desmet G, Jimidar ISM. Triboelectric charging of particles, an ongoing matter: From the early onset of planet formation to assembling crystals. ACS Omega. 2022;7(46):41828–39.

[47] Ohshima H, Furusawa K, editors. Electrical Phenomena at Interfaces: Fundamentals, Measurements, and Applications, 2nd ed. New York: Marcel Dekker, Inc.; 1998.

[48] Ohshima H. Theory of electrostatics and electrokinetics of soft particles. Science and Technology of Advanced Materials. 2009;10(6):063001.

[49] Ashrafizadeh SN, Seifollahi Z, Ganjizade A, Sadeghi A. Electrophoresis of spherical soft particles in electrolyte solutions: A review. Electrophoresis. 2020;41(1–2):81–103.

[50] Gopmandal PP, Duval JFL. Electrostatics and electrophoresis of engineered nanoparticles and particulate environmental contaminants: Beyond zeta potential-based formulation. Current Opinion in Colloid & Interface Science. 2022;60:101605.

[51] Lyklema J. Fundamentals of Interface and Colloid Science. London: Elsevier Academic Press; 2005.

[52] Maurya SK, Gopmandal PP, Ohshima H, Duval JFL. Electrophoresis of composite soft particles with differentiated core and shell permeabilities to ions and fluid flow. Journal of Colloid and Interface Science. 2020;558:280–90.

[53] Ohshima H. Chapter 12 – Electrophoresis: Motion of Colloidal Particles in an Electric Field. In: Ohshima H, editor. Interface Science and Technology, Vol. 37. London: Elsevier; 2024. pp. 281–330.

[54] O'Brien RW, White LR. Electrophoretic mobility of a spherical colloidal particle. Journal of the Chemical Society, Faraday Transactions 2: Molecular and Chemical Physics. 1978;74(0):1607–26.

[55] Ohshima H. Chapter 13 – Electrophoresis of Soft Particles. In: Ohshima H, editor. Interface Science and Technology, Vol. 37. London: Elsevier; 2024. pp. 331–65.

[56] Israelachvili JN. Intermolecular and Surface Forces, 3rd ed. Amsterdam: Elsevier; 2011.

[57] Hiemenz PC, Rajagopalan R. Principles of Colloid and Surface Chemistry, 3rd ed. New York: Marcel Dekker; 1997.

[58] Shaw DJ. Introduction to Colloid and Surface Chemistry, 3rd ed. London: Butterworth & Co. Ltd.; 1980.

[59] Mozaffari Majd M, Kordzadeh-Kermani V, Ghalandari V, Askari A, Sillanpää M. Adsorption isotherm models: A comprehensive and systematic review (2010–2020). Science of the Total Environment. 2022;812:151334.

[60] Lyklema J. Fundamentals of Interface and Colloid Science. London: Academic Press; 1995.

[61] Wang J, Guo X. Adsorption kinetic models: Physical meanings, applications, and solving methods. Journal of Hazardous Materials. 2020;390:122156.

[62] Murphy OP, Vashishtha M, Palanisamy P, Kumar KV. A review on the adsorption isotherms and design calculations for the optimization of adsorbent mass and contact time. ACS Omega. 2023;8 (20):17407–30.

[63] Kontogeorgis GM, Kiil S. Introduction to Applied Colloid and Surface Chemistry. West Sussex: Wiley; 2016.

[64] Labuda J, Barek J, Gajdosechova Z, Jacob S, Johnston L, Krystek P, et al. Analytical chemistry of engineered nanomaterials: Part 2. analysis in complex samples (IUPAC Technical Report). Pure and Applied Chemistry. 2023;95(11):1159–96.

[65] Mourdikoudis S, Pallares RM, Thanh NTK. Characterization techniques for nanoparticles: Comparison and complementarity upon studying nanoparticle properties. Nanoscale. 2018;10 (27):12871–934.

[66] Chestnut MH. Confocal microscopy of colloids. Current Opinion in Colloid & Interface Science. 1997;2(2):158–61.

[67] Gregory DG, Yao N. The Advanced Microscopy of Colloids. In: Priestley R, Prud'homme R, editors. Polymer Colloids: Formation, Characterization and Applications. The Royal Society of Chemistry; 2019.

[68] Cerbino R. Quantitative optical microscopy of colloids: The legacy of Jean Perrin. Current Opinion in Colloid & Interface Science. 2018;34:47–58.

[69] Elliot MS, Poon WCK. Conventional optical microscopy of colloidal suspensions. Advances in Colloid and Interface Science. 2001;92(1):133–94.

[70] Swart I, Liljeroth P, Vanmaekelbergh D. Scanning probe microscopy and spectroscopy of colloidal semiconductor nanocrystals and assembled structures. Chemical Reviews. 2016;116(18):11181–219.

[71] Kappl M, Butt H-J. The colloidal probe technique and its application to adhesion force measurements. Particle & Particle Systems Characterization. 2002;19(3):129–43.

[72] Joshua AM, Cheng G, Lau EV. Soft matter analysis via atomic force microscopy (AFM): A review. Applied Surface Science Advances. 2023;17:100448.

[73] Maurice PA. Applications of atomic-force microscopy in environmental colloid and surface chemistry. Colloids and Surfaces A: Physicochemical and Engineering Aspects. 1996;107:57–75.

[74] Mozhayeva D, Engelhard C. A critical review of single particle inductively coupled plasma mass spectrometry – A step towards an ideal method for nanomaterial characterization. Journal of Analytical Atomic Spectrometry. 2020;35(9):1740–83.

[75] Meng Z, Zheng L, Fang H, Yang P, Wang B, Li L, et al. Single particle inductively coupled plasma time-of-flight mass spectrometry – A powerful tool for the analysis of nanoparticles in the environment. Processes. 2023;11(4):1237.

[76] Pornwilard M-M, Siripinyanond A. Field-flow fractionation with inductively coupled plasma mass spectrometry: Past, present, and future. Journal of Analytical Atomic Spectrometry. 2014;29(10):1739–52.

[77] Pansu M, Gautheyrou J. Fractionation of the Colloidal Systems. In: Handbook of Soil Analysis: Mineralogical, Organic and Inorganic Methods. Berlin, Heidelberg: Springer Berlin Heidelberg; 2006. pp. 65–82.

[78] Baalousha M, Stolpe B, Lead JR. Flow field-flow fractionation for the analysis and characterization of natural colloids and manufactured nanoparticles in environmental systems: A critical review. Journal of Chromatography A. 2011;1218(27):4078–103.

[79] OMLC Web-Based Calculators, https://omlc.org/index.html [Internet]. 2018. Available from: https://omlc.org/index.html.

[80] Sun SF. Physical Chemistry of Macromolecules: Basic Principles and Issues. New York: John Wiley & Sons, Inc.; 1994.

[81] Kong Z, Fei R, Cheng Y, Wang X, Xu N, Wang Z, et al. Imaging-based lidar for atmospheric remote sensing: A review. Optics & Laser Technology. 2025;191:113354.

[82] Barcikowski S, Amendola V, Lau M, Marzun G, Rehbock C, Reichenberger Dr. Rer. Nat S, et al. Handbook of Laser Synthesis & Processing of Colloids, Second edition ed. DuEPublico, University of Duisburg-Essen 2019 September 30.

[83] Gökce B, Amendola V, Barcikowski S. Opportunities and challenges for laser synthesis of colloids. ChemPhysChem. 2017;18(9):983–85.

[84] Gökce B, Rehbock C, Ramesh V, Kohsakowski S, Hupfeld T, Reichenberger S, et al. Laser Synthesis of Colloids: Applications. In: Sugioka K, editor. Handbook of Laser Micro- and Nano-Engineering. Cham: Springer International Publishing; 2020. pp. 1–25.

[85] Zhang D, Gökce B, Barcikowski S. Laser synthesis and processing of colloids: Fundamentals and applications. Chemical Reviews. 2017;117(5):3990–4103.

[86] Theerthagiri J, Karuppasamy K, Lee SJ, Shwetharani R, Kim H-S, Pasha SKK, et al. Fundamentals and comprehensive insights on pulsed laser synthesis of advanced materials for diverse photo- and electrocatalytic applications. Light: Science & Applications. 2022;11(1):250.

[87] McGloin D, McDonald C, Belotti Y. Colloidal interactions with optical fields: Optical tweezers. Chapter 7 in Fluids, Colloids and Soft Materials. 2016;111–30.

[88] Grier DG. Optical tweezers in colloid and interface science. Current Opinion in Colloid & Interface Science. 1997;2(3):264–70.

[89] Boyd RW. Nonlinear Optics. New York: Academic Press; 2003.

[90] Stucky GD, Marder SR, Sohn JE. Linear and Nonlinear Polarizability. In: Materials for Nonlinear Optics. ACS Symposium Series. 455. American Chemical Society; 1991. pp. 2–30.
[91] Shen YR. The Principles of Nonlinear Optics. Oxford: Wiley Classics Library; 2003.
[92] Franken PA, Hill AE, Peters CW, Weinreich G. Generation of optical harmonics. Physical Review Letters. 1961;7(4):118–19.
[93] Mukamel S. Principles of Nonlinear Optical Spectroscopy. New York: Oxford University Press; 1995.
[94] Simpson GJ. Nonlinear Optical Polarization Analysis in Chemistry and Biology. Cambridge: Cambridge University Press; 2017.
[95] Eisenthal KB. Second harmonic spectroscopy of aqueous nano- and microparticle interfaces. Chemical Reviews. 2006;106(4):1462–77.
[96] Ray PC. Size and shape dependent second order nonlinear optical properties of nanomaterials and their application in biological and chemical sensing. Chemical Reviews. 2010;110(9):5332–65.
[97] Roke S, Gonella G. Nonlinear light scattering and spectroscopy of particles and droplets in liquids. Annual Review of Physical Chemistry. 2012;63(1):353–78.
[98] Gonella G, Dai H-L. Second harmonic light scattering from the surface of colloidal objects: Theory and applications. Langmuir. 2014;30(10):2588–99.
[99] Beer A, Roke S, Dadap J. Theory of optical second-harmonic and sum-frequency scattering from arbitrarily shaped particles. Journal of the Optical Society of America B. 2011;28:1374–84.
[100] Wang H, Yan ECY, Borguet E, Eisenthal KB. Second harmonic generation from the surface of centrosymmetric particles in bulk solution. Chemical Physics Letters. 1996;259(1):15–20.
[101] Roke S, Bonn M, Petukhov AV. Nonlinear optical scattering: The concept of effective susceptibility. Physical Review B. 2004;70(11):115106.
[102] Qian Y, Deng G-H, Rao Y. In situ spectroscopic probing of polarity and molecular configuration at aerosol particle surfaces. The Journal of Physical Chemistry Letters. 2020;11(16):6763–71.
[103] Schneider L, Schmid HJ, Peukert W. Influence of particle size and concentration on the second-harmonic signal generated at colloidal surfaces. Applied Physics B. 2007;87(2):333–39.
[104] Laven P. MiePlot – Mie Scattering. 2021. http://www.philiplaven.com/mieplot.htm
[105] Jen S-H, Gonella G, Dai H-L. The effect of particle size in second harmonic generation from the surface of spherical colloidal particles. I: Experimental observations. The Journal of Physical Chemistry A. 2009;113(16):4758–62.
[106] Yang N, Angerer WE, Yodh AG. Angle-resolved second-harmonic light scattering from colloidal particles. Physical Review Letters. 2001;87(10):103902.
[107] Jen S-H, Dai H-L, Gonella G. The effect of particle size in second harmonic generation from the surface of spherical colloidal particles. II: The nonlinear Rayleigh–Gans–Debye model. The Journal of Physical Chemistry C. 2010;114(10):4302–08.
[108] de Beer AGF, Roke S. Nonlinear Mie theory for second-harmonic and sum-frequency scattering. Physical Review B. 2009;79(15):155420.
[109] Wunderlich S, Schürer B, Sauerbeck C, Peukert W, Peschel U. Molecular Mie model for second harmonic generation and sum frequency generation. Physical Review B. 2011;84(23):235403.
[110] de Beer AGF, Roke S. Obtaining molecular orientation from second harmonic and sum frequency scattering experiments in water: Angular distribution and polarization dependence. The Journal of Chemical Physics. 2010;132. 234702.
[111] Ong S, Zhao X, Eisenthal KB. Polarization of water molecules at a charged interface: Second harmonic studies of the silica/water interface. Chemical Physics Letters. 1992;191(3):327–35.
[112] Gonella G, Lütgebaucks C, de Beer AGF, Roke S. Second harmonic and sum-frequency generation from aqueous interfaces is modulated by interference. The Journal of Physical Chemistry C. 2016;120(17):9165–73.
[113] Wen Y-C, Zha S, Liu X, Yang S, Guo P, Shi G, et al. Unveiling microscopic structures of charged water interfaces by surface-specific vibrational spectroscopy. Physical Review Letters. 2016;116(1):016101.

[114] Ohno PE, Wang H-F, Geiger FM. Second-order spectral lineshapes from charged interfaces. Nature Communications. 2017;8(1):1032.
[115] Ohno PE, Saslow SA, Wang H-F, Geiger FM, Eisenthal KB. Phase-referenced nonlinear spectroscopy of the α-quartz/water interface. Nature Communications. 2016;7(1):13587.
[116] Ohno PE, Chang H, Spencer AP, Liu Y, Boamah MD, Wang H-F, et al. Beyond the Gouy–Chapman Model with heterodyne-detected second harmonic generation. The Journal of Physical Chemistry Letters. 2019;10(10):2328–34.
[117] Zhang C. Sum frequency generation vibrational spectroscopy for characterization of buried polymer interfaces. Applied Spectroscopy. 2017;71(8):1717–49.
[118] Wang H, Yan ECY, Liu Y, Eisenthal KB. Energetics and population of molecules at microscopic liquid and solid surfaces. The Journal of Physical Chemistry B. 1998;102(23):4446–50.
[119] Campen RK, Zheng D-S, Wang H-F, Borguet E. Second harmonic generation as a probe of multisite adsorption at solid–liquid interfaces of aqueous colloid suspensions. The Journal of Physical Chemistry C. 2007;111(25):8805–13.
[120] Eckenrode HM, Jen S-H, Han J, Yeh A-G, Dai H-L. Adsorption of a cationic dye molecule on polystyrene microspheres in colloids: Effect of surface charge and composition probed by second harmonic generation. The Journal of Physical Chemistry B. 2005;109(10):4646–53.
[121] Wang H-F, Troxler T, Yeh A-G, Dai H-L. Adsorption at a carbon black microparticle surface in aqueous colloids probed by optical second-harmonic generation. The Journal of Physical Chemistry C. 2007;111(25):8708–15.
[122] Wang H, Troxler T, Yeh A-G, Dai H-L. In situ, nonlinear optical probe of surfactant adsorption on the surface of microparticles in colloids. Langmuir. 2000;16(6):2475–81.
[123] Yamaguchi S, Tahara T. Development of electronic sum frequency generation spectroscopies and their application to liquid interfaces. The Journal of Physical Chemistry C. 2015;119(27):14815–28.
[124] Dhami S, Kumar Y, Pandey R. Development of electronic sum frequency generation spectrophotometer to assess the buried interfaces. Biointerphases. 2023;18(4):041201.
[125] Eckenrode HM, Dai H-L. Nonlinear optical probe of biopolymer adsorption on colloidal particle surface: Poly-l-lysine on polystyrene sulfate microspheres. Langmuir. 2004;20(21):9202–09.
[126] Yan ECY, Liu Y, Eisenthal KB. New method for determination of surface potential of microscopic particles by second harmonic generation. The Journal of Physical Chemistry B. 1998;102(33):6331–36.
[127] Sara RJ, Coers D, Behrman C, Bobay J, Subir M. Molecular adsorption and physicochemical properties at liquid/liquid nanoemulsion soft interfaces: Effect of charge and hydrophobicity. Journal of Physical Chemistry B. 2024;128(12):3004–15.
[128] Xu B, Gonella G, DeLacy BG, Dai H-L. Adsorption of anionic thiols on silver nanoparticles. The Journal of Physical Chemistry C. 2015;119(10):5454–61.
[129] Wu Y, Li W, Xu B, Li X, Wang H, McNeill VF, et al. Observation of organic molecules at the aerosol surface. The Journal of Physical Chemistry Letters. 2016;7(12):2294–97.
[130] Kanu I, Ojha A, Adams K, Brooks J, Subir M. Surface functional group dependent enthalpic and entropic contributions to molecular adsorption on colloidal microplastics. Soft Matter. 2025;21(34):6814–22.
[131] Karam TE, Haber LH. Molecular adsorption and resonance coupling at the colloidal gold nanoparticle interface. The Journal of Physical Chemistry C. 2014;118(1):642–49.
[132] Gan W, Xu B, Dai H-L. Super bright luminescent metallic nanoparticles. The Journal of Physical Chemistry Letters. 2018;9(15):4155–59.
[133] Gan W, Gonella G, Zhang M, Dai H-L. Communication: Reactions and adsorption at the surface of silver nanoparticles probed by second harmonic generation. The Journal of Chemical Physics. 2011;134(4). 041104.

[134] You Y, Bloomfield A, Liu J, Fu L, Herzon SB, Yan ECY. Real-time kinetics of surfactant molecule transfer between emulsion particles probed by in situ second harmonic generation spectroscopy. Journal of the American Chemical Society. 2012;134(9):4264–68.
[135] Ranasinghe JC, Dikkumbura AS, Hamal P, Chen M, Khoury RA, Smith HT, et al. Monitoring the growth dynamics of colloidal gold-silver core-shell nanoparticles using in situ second harmonic generation and extinction spectroscopy. The Journal of Chemical Physics. 2019;151(22):224701.
[136] Srivastava A, Eisenthal KB. Kinetics of molecular transport across a liposome bilayer. Chemical Physics Letters. 1998;292(3):345–51.
[137] Yan EC, Eisenthal KB. Effect of cholesterol on molecular transport of organic cations across liposome bilayers probed by second harmonic generation. Biophysical Journal. 2000;79(2):898–903.
[138] Liu J, Subir M, Nguyen K, Eisenthal KB. Second harmonic studies of ions crossing liposome membranes in real time. Journal of Physical Chemistry B. 2008;112(48):15263–66.
[139] Zeng J, Eckenrode Heather M, Dounce Susan M, Dai H-L. Time-resolved molecular transport across living cell membranes. Biophysical Journal. 2013;104(1):139–45.
[140] Wilhelm MJ, Sharifian Gh M, Dai H-L. Chemically induced changes to membrane permeability in living cells probed with nonlinear light scattering. Biochemistry. 2015;54(29):4427–30.
[141] Wilhelm MJ, Sharifian Gh M, Wu T, Li Y, Chang CM, Ma J, et al. Determination of bacterial surface charge density via saturation of adsorbed ions. Biophysical Journal. 2021;120(12):2461–70.
[142] Miller LN, Brewer WT, Williams JD, Fozo EM, Calhoun TR. Second harmonic generation spectroscopy of membrane probe dynamics in gram-positive bacteria. Biophysical Journal. 2019;117(8):1419–28.
[143] Page EF, Blake MJ, Foley GA, Calhoun TR. Monitoring membranes: The exploration of biological bilayers with second harmonic generation. Chemical Physics Reviews. 2022;3(4):041307.
[144] Chen S, Hou Y, Xu B, Chen S, Yuan Q, Gan W. On the models for understanding the second harmonic generation scattering from lipid membranes and the corresponding molecular kinetics. Colloids and Surfaces A: Physicochemical and Engineering Aspects. 2024;683:133004.
[145] Wilhelm MJ, Dai H-L. Molecule-membrane interactions in biological cells studied with second harmonic light scattering. Chemistry – An Asian Journal. 2020;15(2):200–13.
[146] Liu Y, Dadap JI, Zimdars D, Eisenthal KB. Study of interfacial charge-transfer complex on TiO2 particles in aqueous suspension by second-harmonic generation. The Journal of Physical Chemistry B. 1999;103(13):2480–86.
[147] Kaylor L, Skelly P, Alsarrani M, Subir M. Enhanced malachite green photolysis at the colloidal-aqueous interface. Chemosphere. 2022;287. 131953.
[148] Joret M, Idlahcen S, Mazur M, Yon J. Non-linear optics for an online probing of the specific surface area of nanoparticles in the aerosol phase. Journal of Aerosol Science. 2025;183:106484.
[149] Nigam S, Rutan S. Principles and applications of solvatochromism. Applied Spectroscopy. 2001;55 (11). 362A – 370A
[150] Reichardt C. Solvatochromic dyes as solvent polarity indicators. Chemical Reviews. 1994;94 (8):2319–58.
[151] Wang H, Borguet E, Eisenthal KB. Polarity of liquid interfaces by second harmonic generation spectroscopy. The Journal of Physical Chemistry A. 1997;101(4):713–18.
[152] Sen S, Yamaguchi S, Tahara T. different molecules experience different polarities at the air/water interface. Angewandte Chemie International Edition. 2009;48(35):6439–42.
[153] Ma G, Allen HC. Diffuse reflection broad bandwidth sum frequency generation from particle surfaces. Journal of the American Chemical Society. 2002;124(32):9374–75.
[154] Bordenyuk AN, Weeraman C, Yatawara A, Jayathilake HD, Stiopkin I, Liu Y, et al. Vibrational sum frequency generation spectroscopy of dodecanethiol on metal nanoparticles. The Journal of Physical Chemistry C. 2007;111(25):8925–33.

[155] Fankhauser AM, Butman JL, Cooke ME, Fyodorova Y, Liu Y, O'Brien RE, et al. Indoor surface chemistry variability: Microspectroscopic analysis of deposited particles in dwellings across the United States. Environmental Science: Processes & Impacts. 2025;27(6):1704–13.
[156] Ebben CJ, Ault AP, Ruppel MJ, Ryder OS, Bertram TH, Grassian VH, et al. Size-resolved sea spray aerosol particles studied by vibrational sum frequency generation. The Journal of Physical Chemistry A. 2013;117(30):6589–601.
[157] Schaller RD, Saykally RJ. Near-field infrared sum-frequency generation imaging of chemical vapor deposited zinc selenide. Langmuir. 2001;17(7):2055–58.
[158] Roke S, Roeterdink WG, Wijnhoven JEGJ, Petukhov AV, Kleyn AW, Bonn M. Vibrational sum frequency scattering from a submicron suspension. Physical Review Letters. 2003;91(25):258302.
[159] Mapile AN, LeRoy MA, Fabrizio K, Scatena LF, Brozek CK. The surface of colloidal metal–organic framework nanoparticles revealed by vibrational sum frequency scattering spectroscopy. ACS Nano. 2024;18(20):13406–14.
[160] Johansson PK, Castner DG. Vibrational sum-frequency scattering as a sensitive approach to detect structural changes in collagen fibers treated with surfactants. Langmuir. 2019;35(24):7848–57. 011201
[161] Schmüser L, Golbek TW, Weidner T. Windowless detection geometry for sum frequency scattering spectroscopy in the C–D and amide I regions. Biointerphases. 2021;16(1):011201.
[162] Rehl B, Ronceray N, Zhang L, Radenovic A, Roke S. Sum-frequency scattering spectroscopy reveals the charging mechanism and surface structure of hBN nanoflakes in solution. ACS Nano. 2025;19 (27):24912–19.
[163] Johansson PK, Koelsch P. Vibrational sum-frequency scattering for detailed studies of collagen fibers in aqueous environments. Journal of the American Chemical Society. 2014;136(39):13598–601.
[164] Pullanchery S, Kulik S, Rehl B, Hassanali A, Roke S. Charge transfer across C–H···O hydrogen bonds stabilizes oil droplets in water. Science. 2021;374(6573):1366–70.
[165] Sang Y, Yang F, Chen S, Xu H, Zhang S, Yuan Q, et al. Molecular interactions at the hexadecane/water interface in the presence of surfactants studied with second harmonic generation. The Journal of Chemical Physics. 2015;142(22):224704.
[166] Zhang C, Gao J, Hankett J, Varanasi P, Borst J, Shirazi Y, et al. Corn oil–water separation: Interactions of proteins and surfactants at corn oil/water interfaces. Langmuir. 2020;36(15):4044–54.
[167] Gan W, Wu W, Yang F, Hu D, Fang H, Lan Z, et al. The behavior of hydroxide and hydronium ions at the hexadecane–water interface studied with second harmonic generation and zeta potential measurements. Soft Matter. 2017;13(43):7962–68.
[168] Saykally RJ. Two sides of the acid–base story. Nature Chemistry. 2013;5(2):82–84.
[169] Vácha R, Rick SW, Jungwirth P, de Beer AGF, de Aguiar HB, Samson J-S, et al. The orientation and charge of water at the hydrophobic oil droplet–water interface. Journal of the American Chemical Society. 2011;133(26):10204–10.
[170] Vácha R, Buch V, Milet A, Devlin JP, Jungwirth P. Autoionization at the surface of neat water: Is the top layer pH neutral, basic, or acidic?. Physical Chemistry Chemical Physics. 2007;9(34):4736–47.
[171] Buch V, Milet A, Vácha R, Jungwirth P, Devlin JP. Water surface is acidic. Proceedings of the National Academy of Sciences. 2007;104(18):7342–47.
[172] El-Zohry AM, Sølling T. Mini-review on water–oil emulsions through vibrational sum frequency generation. Energy & Fuels. 2024;38(2):743–60.
[173] Hosseinpour S, Götz V, Peukert W. Effect of surfactants on the molecular structure of the buried oil/water interface. Angewandte Chemie International Edition. 2021;60(47):25143–50.
[174] Carpenter AP, Tran E, Altman RM, Richmond GL. Formation and surface-stabilizing contributions to bare nanoemulsions created with negligible surface charge. Proceedings of the National Academy of Sciences. 2019;116(19):9214–19.

[175] Hensel JK, Carpenter AP, Ciszewski RK, Schabes BK, Kittredge CT, Moore FG, et al. Molecular characterization of water and surfactant AOT at nanoemulsion surfaces. Proceedings of the National Academy of Sciences. 2017;114(51):13351–56.
[176] Pullanchery S, Kulik S, Roke S. Water structure at the hydrophobic nanodroplet surface revealed by vibrational sum frequency scattering using isotopic dilution. The Journal of Physical Chemistry B. 2022;126(16):3186–92.
[177] Zhang L, Pullanchery S, Cremer PS, Roke S. Finite volume effects in water nanodroplets: A molecular level investigation. ACS Nano. 2025;19(26):23829–39.
[178] de Aguiar HB, de Beer AGF, Strader ML, Roke S. The interfacial tension of nanoscopic oil droplets in water is hardly affected by SDS surfactant. Journal of the American Chemical Society. 2010;132 (7):2122–23.
[179] de Aguiar HB, Strader ML, de Beer AGF, Roke S. Surface structure of sodium dodecyl sulfate surfactant and oil at the oil-in-water droplet liquid/liquid interface: A manifestation of a nonequilibrium surface state. The Journal of Physical Chemistry B. 2011;115(12):2970–78.
[180] Zdrali E, Chen Y, Okur HI, Wilkins DM, Roke S. The molecular mechanism of nanodroplet stability. ACS Nano. 2017;11(12):12111–20.
[181] Subir M, Liu J, Eisenthal KB. Protonation at the aqueous interface of polymer nanoparticles with second harmonic generation (vol 112, pg 15809, 2008). The Journal of Physical Chemistry C. 2009;113(48):20772.
[182] Glikman D, Braunschweig B. Nanoscale effects on the surfactant adsorption and interface charging in hexadecane/water emulsions. ACS Nano. 2021;15(12):20136–47.
[183] Petersen PB, Saykally RJ. Is the liquid water surface basic or acidic? Macroscopic vs. molecular-scale investigations. Chemical Physics Letters. 2008;458(4):255–61.
[184] Beattie JK, Djerdjev AM, Warr GG. The surface of neat water is basic. Faraday Discussions. 2009;141 (0):31–39.
[185] Beattie JK, Djerdjev AM. The pristine oil/water interface: Surfactant-free hydroxide-charged emulsions. Angewandte Chemie International Edition. 2004;43(27):3568–71.
[186] Zimmermann R, Freudenberg U, Schweiß R, Küttner D, Werner C. Hydroxide and hydronium ion adsorption – A survey. Current Opinion in Colloid & Interface Science. 2010;15(3):196–202.
[187] Zhang P, Feng M, Xu X. Double-layer distribution of hydronium and hydroxide ions in the air–water interface. ACS Physical Chemistry Au. 2024;4(4):336–46.
[188] Yamaguchi S, Kundu A, Sen P, Tahara T. Communication: Quantitative estimate of the water surface pH using heterodyne-detected electronic sum frequency generation. The Journal of Chemical Physics. 2012;137(15):151101.
[189] Qian Y, Brown JB, Huang-Fu Z-C, Zhang T, Wang H, Wang S, et al. In situ analysis of the bulk and surface chemical compositions of organic aerosol particles. Communications Chemistry. 2022;5(1):58.
[190] Qian Y, Brown JB, Huang-Fu Z-C, Zhang T, Wang H, Wang S, et al. Reply to: A discrepancy of 107 in experimental and theoretical density detection limits of aerosol particles by surface nonlinear light scattering. Communications Chemistry. 2023;6(1):115.
[191] Marchioro A, Golbek TW, Chatterley AS, Weidner T, Roke S. A discrepancy of 107 in experimental and theoretical density detection limits of aerosol particles by surface nonlinear light scattering. Communications Chemistry. 2023;6(1):114.
[192] Mishchenko MI, Hovenier JW, Travis LD. Light Scattering by Nonspherical Particles: Theory, Measurements, and Applications. New York: Academic Press; 2000.
[193] Yamaguchi S, Otosu T. Progress in phase-sensitive sum frequency generation spectroscopy. Physical Chemistry Chemical Physics. 2021;23(34):18253–67.
[194] Shen YR. Phase-sensitive sum-frequency spectroscopy. Annual Review of Physical Chemistry. 2013;64:129–50. Volume 64, 2013.

[195] Fellows AP, Duque ÁD, Balos V, Lehmann L, Netz RR, Wolf M, et al. Sum-frequency generation spectroscopy of aqueous interfaces: The role of depth and its impact on spectral interpretation. The Journal of Physical Chemistry C. 2024;128(49):20733–50.

[196] Nihonyanagi S, Yamaguchi S, Tahara T. Ultrafast dynamics at water interfaces studied by vibrational sum frequency generation spectroscopy. Chemical Reviews. 2017;117(16):10665–93.

[197] Saak C-M, Backus EHG. The role of sum-frequency generation spectroscopy in understanding on-surface reactions and dynamics in atmospheric model-systems. The Journal of Physical Chemistry Letters. 2024;15(17):4546–59.

[198] Deng G-H, Qian Y, Rao Y. Development of ultrafast broadband electronic sum frequency generation for charge dynamics at surfaces and interfaces. The Journal of Chemical Physics. 2019;150(2): 024708.

[199] Benderskii A, Morita A. Nonlinear spectroscopy and interfacial structure and dynamics. The Journal of Chemical Physics. 2019;151(15):150401.

[200] Deng G-H, Wei Q, Qian Y, Zhang T, Leng X, Rao Y. Development of interface-/surface-specific two-dimensional electronic spectroscopy. Review of Scientific Instruments. 2021;92(2):023104.

[201] Huang-Fu ZC, Qian Y, Deng GH, Zhang T, Schmidt S, Brown J, et al. Development of two-dimensional electronic-vibrational sum frequency generation (2D-EVSFG) for vibronic and solvent couplings of molecules at interfaces and surfaces. ACS Physical Chemistry Au. 2023;3(4):374–85.

Index

 | https://doi.org/10.1515/9783111018386-013

Printed and bound by CPI Group (UK) Ltd, Croydon, CR0 4YY
13/09/2026
14960853-0001